Exchanging Data from SAS® to Excel

The ODS Excel Destination

William E. Benjamin, Jr.

S.sas.

sas.com/books

The correct bibliographic citation for this manual is as follows: Benjamin, William E., Jr. 2017. *Exchanging Data From SAS® to Excel: The ODS Excel Destination.* Cary, NC: SAS Institute Inc.

Exchanging Data From SAS® to Excel: The ODS Excel Destination

Copyright © 2017, SAS Institute Inc., Cary, NC, USA

ISBN 978-1-62960-609-5 (Hard copy)
ISBN : 978-1-63526-140-0 (EPUB)
ISBN 978-1-63526-141-7 (MOBI)
ISBN 978-1-63526-142-4 (PDF)

Contents

About This Book

What Does This Book Cover?

I wrote this book to help SAS users of all skill levels learn to use the new SAS ODS EXCEL Destination software. My years of programming experience have helped me decode the mysteries of vendor-supplied system documentation. I wanted to convert that information into practical examples of how to apply the options to every day programming applications. Microsoft Excel is one of the most widely used software tools available to computer users. While many computer programmers are also Excel software users, there are far more Excel users than programmers that use Excel. Of course, this book is for SAS users who move data to Excel workbooks. The intent of this book is to enable SAS users to format the Excel Workbook output in a way that eliminates manual changes after the workbook is created.

This book explains how to use all of the options in the ODS EXCEL destination available in SAS 9.4 (TS1M3). I have broken up the options and suboptions into groups that I feel work on similar parts of the Excel output workbooks. My grouping is different from any other grouping that is provided by SAS Institute or anyone else.

This book uses only the Microsoft Windows operating system version of Base SAS to execute the examples. However, since the ODS Excel destination is a supported part of Base SAS, the code and examples will work on other operating systems (IBM Mainframes, UNIX, Linux, and perhaps others.) In addition, with the proper library assignments this will work on other SAS Products, like SAS Enterprise Guide, SAS Studio, and SAS University Edition.

Is This Book for You?

Whatever your skill level, I hope you will find examples that will teach you something. In every class I teach or paper I present, I always ask whether anyone learned anything. I want you to be able to find a place on your desk for this book, use it as you progress through the skills presented, and gain expertise to easily move your data.

What Are the Prerequisites for This Book?

This book is designed for you to use without the need for prerequisites. If you can open the SAS program and copy data using your mouse, then you can get started. I do not attempt to teach you how to write SAS programs or build an Excel spreadsheet, but I present methods to move data between the two data storage tools.

What Should You Know about the Examples?

This book includes software examples for you to follow to gain hands-on experience with SAS.

Software Used to Develop the Book's Content

SAS 9.4 (TS1M3) was used was used to create all the examples in this book. I used Microsoft Excel 2013 while writing this book. I also know that the files generated are compatible with Microsoft Excel 2016. I expect them to be compatible with Microsoft Excel 2007 and Excel 2010. The output files are always in the new Excel file format (*.xlsx).

Example Code and Data

The example code shown in this book was executed using the Base SAS Display Manager on a Microsoft Windows 10 Operating System. I expect it to execute in any other Base SAS environment. Most of the examples use a subset of the SASHELP.SHOES data set that I called ASIA_ONLY, because the name implies that it includes only the data from the region "ASIA" of the SASHELP.SHOES data set. Some examples use the whole SASHELP.SHOES data set, and a few others are identified when they are used.

You can access the example code and data for this book by linking to its author page at https://support.sas.com/authors. Each chapter has a segment of code that is not shown in the book (except for Chapter 2), which creates a path and sets some other values. The code name has a general format of "SAS Code 3.0 Create ASIA_ONLY file in Chapter 03 Path.sas," You will need to modify these modules to execute the code.

SAS University Edition

If you are using SAS University Edition to access data and run your programs, then please check the SAS University Edition page to ensure that the software contains the product or products that you need to run the code. The link is www.sas.com/universityedition.

Because the SAS University Edition run-time environment differs slightly from other SAS products, I found that special code was needed that might not be needed in SAS Studio. I recommend using SAS University Edition because it is a great way for new SAS users to learn how to use SAS software without spending a lot of money. Please see the following paper I wrote, available at the following URL: https://pharmasug.org/proceedings/2017/AD/PharmaSUG-2017-AD12.pdf . This paper was presented at PharmaSUG 2017 and is titled "Using the ODS EXCEL Destination with SAS® University Edition to Send Graphs to Excel."

Output and Graphics

All output and graphs were generated by using Base SAS or SAS/GRAPH. The code can be found in the book or on my SAS Author page, located at the following URL: https://support.sas.com/authors.

We Want to Hear from You

SAS Press books are written *by* SAS Users *for* SAS Users. We welcome your participation in their development and your feedback on SAS Press books that you are using. Please visit sas.com/books to do the following:

- Sign up to review a book
- Recommend a topic

- Request information about how to become a SAS Press author
- Provide feedback on a book

Do you have questions about a SAS Press book that you are reading? Contact the author through saspress@sas.com or https://support.sas.com/author_feedback.

SAS has many resources to help you find answers and expand your knowledge. If you need additional help, see our list of resources located at the following URL: sas.com/books.

x

About The Author

William E. Benjamin, Jr., owns Owl Computer Consultancy, LLC, and works as a consultant, trainer, and author. William has been a SAS user for over 30 years and a consultant since 2007. He received an MBA from Western International University and a BS in computer science from Arizona State University. He has written and presented papers for SAS Global Forum, as well as many regional and local SAS users groups.

Learn more about this author by visiting his author page at http://support.sas.com/benjamin. There you can download free book excerpts, access example code and data, read the latest reviews, get updates, and more.

Chapter 1: Introduction

Introduction

Since I began writing code in 1973 to make computers do what I wanted them to do, instead of what they wanted to do, I have always worked as an application programmer. That means that I used someone's software to build a tool to do something. When I upgrade my software the tools I have already written still need to work, because not every computer is upgraded at the same time and new features need to be available to everyone.

While Base SAS and other features of SAS software make SAS an important and useful software suite, many people are not programmers or analysts and do not write computer programs. These people have learned to use and understand how Microsoft Excel can benefit them. They are quite often managers, directors, and vice presidents, and they want to use the data in a comfortable format. Since the release of Microsoft Excel 2007 and the Open XML format, the upgrades made to Excel and computer hardware in general have caused many updates and changes to the way SAS processes Microsoft Excel workbooks and worksheets.

The first adaptations came in the form of learning to read and write the new Excel formats. These changes upgraded PROC EXPORT, PROC IMPORT, the SAS LIBNAME statement, and the new SAS PC Files Server. The introduction of the SAS PC Files Server bridged the 32 to 64-bit barrier in application software, operating system software, and computer hardware. The release of the SAS ODS EXCELXP tagset occurred before the introduction of Excel 2007. The EXCELXP tagset writes output text files in the Extensible Markup Language (XML) format. Both Excel 2003 and 2007 can read these files. The new EXCEL 2007 (and later) workbooks are stored in the open XML format. Between the release of the EXCELXP tagset and December 2014, over 100 modifications in several releases occurred.

In late December 2014, I found out about a new SAS ODS feature, the ODS EXCEL destination. After nearly eight years of research and preparation my first book was days away from being finalized for publication. I realized I had the software required to create an example and lobbied hard to get something into my first book. This second book replaces the page and a half that introduced the ODS EXCEL destination in the first book.

What is the SAS ODS EXCEL Destination?

The SAS Output Delivery System has two major features that deal with outputting data files that Microsoft Excel can read. The general classifications are ODS tagsets and ODS destinations. Each of these has many features. Tagsets are created by PROC TEMPLATE code that any SAS programmer can modify, while the ODS destinations are fixed executable modules that the user cannot modify. While the ODS EXCEL destination is a new part of Base SAS, many of the features of the ODS EXCELXP tagset are available to the new ODS EXCEL destination. See Chapter 2, "ODS Tagset versus Destination" for more information.

Why I like Backward Compatibility

Backward compatibility is a challenge that nearly every software producer faces. It does not matter if you develop software suites, operating systems, or reports for your boss. The only thing you can guarantee about the software, data, or reports that you need to produce is that they will change. SAS is one of the software producers that do a good job of maintaining backward compatibility. Sometimes I recode legacy features to accept the same inputs as before but I also accept new options and output new information. This allows the legacy code to run seamlessly but give new features to existing routines. At other times the SAS documentation is updated to say that an option is retained for backward compatibility, but is not used by the software.

I like backward capability because it enables me to write new code inside a current feature of a tool I have written. This leaves the existing functions intact and lets me write new features. Thus, allowing me to have one upgraded program without losing features.

Why I Cannot Use Backward Compatibility for This Book

The change from the binary formatted *.xls file structure to the open XML file format was a big file format change that was introduced with Microsoft Office 2007. The new output *.xlsx file format, which included "zip" files of XML code, also changed the access method for all new Excel workbook files. Any SAS user that needed to read or write an Excel worksheet needed to consider the Excel workbook format changes. Everyone who had used or programmed applications using the old formats suddenly had to redo every application that used the old formats.

SAS debuted the Output Delivery System (ODS) with Version 7 of Base SAS. SAS Versions 8 and 9 expanded upon and introduced methods that would allow data transfers between SAS and Microsoft Excel. The transfer methods called destinations (like CSV or HTML) or tagsets (like CSV, HTML, MSOFFICE2K, or EXCELXP) allowed SAS code to create files that Microsoft Excel could read. After the release of Microsoft Office 2007, SAS wrote many versions of several ODS tools to address the changes to the Microsoft Office products. The ODS tagsets related to files Microsoft Excel could read were all created as text files that Excel could read, but did not directly write to the new Excel file open XML format.

A tagset is a module of SAS template code that PROC TEMPLATE compiles and stores in the SASHELP library with each installed version of SAS. Every user with access to the SASHELP library can use these tagsets. A SAS user can apply changes to a tagset (or create one of their own) and compile this to a user library. This is usually accessible to only one computer, but the user tagsets are sharable.

The new SAS ODS EXCEL destination (not a tagset) is unique in that it creates a file in the Open XML Format, regardless of the filename requested. Providing an output filename with an extension of "*.xls" does not create a binary formatted Excel file that Excel 2003 can read. However, when using the "*.xls" extension as an output Excel workbook name Excel will send an error message and might (depending on the Excel version) enable you to open the created file if you continue.

There is no history for this feature to be backward compatible with. However, some features of the ODS EXCEL destination are similar to features of the SAS ODS tagset EXCELXP. While I did not provide details for every feature of the EXCELXP tagset in my previous book, I did list most of the features that are available. Most of the features that were added with EXCELXP are available with the ODS Excel destination.

Why the ODS Excel Destination Is an Output-Only Feature of SAS

By definition, the SAS ODS (Output Delivery System) features are "OUTPUT" only. Once the file is defined with the ODS FILE statement, created with SAS code, and closed with the ODS <destination> CLOSE statement it is available for use by other programs or SAS features. If the SAS/ACCESS for PC Files software is installed, then you can use the LIBNAME statement to open a new Excel Workbook.

Layout of This Book

The intent of this book is to describe in detail the options of the ODS EXCEL destination. I will be grouping the arguments and options into what I consider common usage elements. I have been an applications programmer for over 40 years and am using my experience to establish what I consider logical units. You should be able to use the index to look up the options alphabetically. I hope my grouping is satisfactory for other needs. While the EXCELXP tagset and the ODS EXCEL Destination are separate and unique features of SAS, many of the options of each tool provide similar output to an output Excel workbook. The following chapters will describe the arguments and options while providing as many examples as space permits. The goal is to show something for everything. The lists below show my general grouping categories. I will describe each of these arguments and options later.

ODS Destination EXCEL Arguments

- File identification
- Excel file properties
- Output features

ODS Destination EXCEL Option Groups

- Workbook
- Worksheet
- Print
- Column
- Row
- Cell level

General Chapter Layout

Chapters 3 and beyond will each introduce and explain some features of the ODS EXCEL destination by providing a description of the feature followed by the SAS code and Excel output. To conserve space each example might contain one or more features. As noted before, the groupings of the features are my own and might not be the same as anyone else might describe them.

Chapter 2: ODS Tagset versus Destination

View of Microsoft Excel Changes from the Outside

During the decade between 2006 and 2015, there have been many changes in the way SAS interfaced with Microsoft Excel. The changes began with the Microsoft release of Office 2007. Because this book deals mainly with Excel, I will refer to that product specifically. For many years the format of the Microsoft Excel files had remained the same. Opening a small Excel file with the Notepad program enabled you to see the binary patterns of the file. While this was generally not a useful thing to do, the patterns did occasionally show readable text. With Microsoft Excel 2007 the Excel file format converted to a compressed file (essentially a ZIP file) of XML files. I am not going to try to decode those files or explain them, it just worked for Microsoft. They could protect their Excel files from having Visual Basic for Applications (VBA) macros inside the *.xlsx files. This allowed users to know there were no macros and the files were safer. Microsoft did allow the Excel files with other file extensions to contain macros and binary components (*.xlsm or *.xlsb file extensions).

However, this provided a problem for everyone else. Many of the older products that were installed could not even open the new Excel files. Microsoft handed out free software to read these new file formats. SAS worked hard to convert their input and output routines to both read the new Excel formats and write the output files that Microsoft Excel could read. A compounding issue over that time period was the advent of new computer chips for the commonly used microprocessors. I still remember my first computer was based on an 8-bit instruction set to make the computer run. It of course had only one Central Processing Unit (CPU) in the approximately 2.5X2.5 inch chipset. It also needed a second chip that size to do the floating-point arithmetic. 64-bit computers started showing up as Personal Computers (PCs) around 2003 [1], and by the time Microsoft Excel 2007 was introduced these computers were well on their way to being the standard for PC computer architecture. The last part of this puzzle is that computer operating system software is generally written to be forward compatible with one architecture. However, the operating systems can allow the program to choose an execution mode (32-bit or 64-bit mode). The problem then became an issue of the data, computer instruction, and memory addressing configurations. Backward compatibility required something to address these issues.

Microsoft Excel had backward compatible features that allowed it to read formatted text files. SAS had PROC TEMPLATE and DATA step features that would allow text files that Microsoft Excel was able to read to be created, either by using DATA step output code or ODS features. The best known and most widely used of these features was the ODS tagset EXCELXP. This tagset produces a specific type of XML, the XML Spreadsheet 2003 format (XMLSS). Excel can read the XMLSS format as a worksheet. SAS users can download the EXCELXP PROC TEMPLATE code and modify the PROC TEMPLATE code to build the tagset. Then they can compile them into their system (starting in version 9.4 the EXCELXP tagset is also pre-loaded into Base SAS). The EXCELXP tagset has many options and features that I discussed in my first book. SAS also provided access to Excel workbooks with the LIBNAME statement, which I will not discuss here. This feature requires SAS/ACCESS for PC Files software.

[1] https://en.wikipedia.org/wiki/64-bit_computing

Data Sources Used as Examples

I like to use data files in the SASHELP directory. Using these files allows everyone to reproduce the examples and then extend the examples to your data files. Because of the nature of some of the examples, several files can be used to produce graphs and other output to more fully explain the options that are available to the users of the ODS EXCEL destination. While I will attempt to explain all of the options, I will at times show several features in the same example.

In an effort to keep all of my examples as close to one page as possible I will be using the following code to produce a SAS data set with only 14 observations. I have provided SAS code on the SAS Authors page labeled as SAS Code (chapter number)-0.sas for each chapter. The code is similar to the code in SAS Code – 2-1 below. You will need to modify the code to match the directories on your computer. In other chapters I will be using the SAS data set "ASIA_ONLY" without referring to this code.

SAS Code 2-1 Create ASIA_ONLY SAS Data Set

```
/***************************************************/
/**  Code to create a SAS work data set ASIA_ONLY  **/
/***************************************************/

data ASIA_ONLY;
    set sashelp.shoes (where=(region="Asia"));
run;

* Create Your Own path name here;
%let path = I:\SAS__Book_2016\Chapter 2 Excel Output;
options nodate;
```

The EXCELXP Tagset

The SAS ODS tagset called EXCELXP is a versatile tagset with many features and options. You can create Excel readable worksheets that have many preset options and features like landscape printing, variable cell sizes, and internal formatting that could take a long time to produce manually. This tagset has over 75 options that can be applied to an output file, and the tagset can actually be modified and stored by SAS users for personal use. The output files produced are commands that Microsoft Excel can read. But if you write the output file using a filename extension that ends in xlsx, Microsoft Excel will give you an error message about the format of the input file. One limitation of the EXCELXP tagset is the inability to produce graphic output in an Excel worksheet directly.

The ODS EXCEL Destination

The newest addition to the way that SAS creates Excel workbooks was released as an experimental version with SAS 9.4 (TS1M1), and is called the SAS ODS EXCEL destination. SAS 9.4 (TS1M3) now has the fully supported version of the ODS Excel destination and it writes Excel workbooks in the native Excel format (Open XML with the .xlsx extension). As a "destination" this ODS feature cannot be modified by a SAS user like the tagset EXCELXP. But that does not mean that the output Excel workbook cannot be modified. The intent of this book is to explain and show you how to use options in the destination to create your own customized workbooks.

Features of the ODS EXCEL Destination

Simple Syntax

The general syntax of the ODS EXCEL statement is rather simple, but the options are many and varied. Note that this book discusses "Actions", "Options", and "Suboptions." For the most complete and updated list of ODS EXCEL options, refer to the ODS documentation at the following URL:
http://support.sas.com/documentation/cdl/en/odsug/69832/PDF/default/odsug.pdf. See Chapter 6 for detailed information and examples of the ODS Excel destination.

SAS Code 2-2 – Simple ODS Excel Syntax

```
Syntax for the ODS EXCEL destination.

ODS EXCEL <(<ID=> identifier)> < action> ;
ODS EXCEL <(<ID=> identifier)> <option(s)> ;
```

The SAS ODS Excel destination syntax shown in SAS Code 2-2 is just the tip of the iceberg. As shown, everything except "ODS EXCEL;" is optional. While the features of the syntax will be explained in later chapters, I will take a moment here to explain the relationships of these syntax elements.

In this book I make a distinction between the "Action" and "Option" features of the ODS Excel destination. However, one thing to point out is that there is an "Argument" called "OPTIONS" that has many "SUB-OPTIONS." I will generally refer to the distinct types as "Actions" or "Additional Arguments" where the "Additional Arguments" are either options or sub-options, as they are described in the SAS HELP under the Base SAS 9.4 (TS1M3) topic "ODS EXCEL Statement." I will clarify this in later chapters.

ID=

This feature enables you to open multiple Excel workbooks during your processing.

Actions

The number of actions is relatively small and deals with generating the output for the Excel workbook. These actions control opening and closing the Excel workbook and selecting, excluding, and listing the SAS objects that are output. These are more fully described in the online documentation at the following URL:
http://support.sas.com/documentation/cdl/en/odsug/69832/PDF/default/odsug.pdf. But an explanation is included in later chapters.

Table 2-1 List of ODS Excel Actions

Actions Include	
NONE	Sends Excel output to the SAS Default output directory.Depending onn your version of SAS, the default directory is shown in the bottom left or right side of the display manager window.
CLOSE	Closes an ODS EXCEL statement with or without an ID= option.
EXCLUDE	An ODS EXCLUDE statement prevents an ODS object from being output.
SELECT	An ODS SELECT statement includes an ODS object in the output.
SHOW	An ODS SHOW statement writes the current selection or exclusion list to the log

Optional Arguments (Options)

The following ODS EXCEL Statement options accept parameters, except the GFOOTNOTE/NOGFOOTNOTE and GTITLE/NOGTITLE option pairs. The OPTIONS option syntax is "OPTIONS(sub-option1 sub-option2 … sub-optionN);". A list of the suboptions is in Table 2-3.

Table 2-2 List of ODS Excel Options

Optional Arguments		
ANCHOR=	AUTHOR=	BOX_SIZING=
CATEGORY=	COMMENTS=	CSSSTYLE=
DOM=	DPI=	FILE=
GFOOTNOTE	NOGFOOTNOTE	GTITLE
NOGTITLE	IMAGE_DPI=	KEYWORDS=
ID=	OPTIONS	SASDATE
STATUS=	STYLE=	TEXT=
TITLE=	WORK=	

These optional arguments will be explained in later chapters.

Suboptions of the OPTIONS Statement

All of these suboptions are accepted on the OPTIONS(…) argument of the ODS EXCEL
statement. Throughout the book I may call them options or suboptions but they are really
suboptions of the OPTIONS option.

Table 2-3 Suboptions of the OPTIONS option of the ODS EXCEL statement

Suboptions of the OPTIONS Argument	
ABSOLUTE_COLUMN_WIDTH=	ABSOLUTE_ROW_HEIGHT=
AUTOFILTER=	BLACKANDWHITE=
BLANK_SHEET=	CENTER_HORIZONTAL=
CENTER_VERTICAL=	COLUMN_REPEAT=
CONTENTS=	DPI=
DRAFTQUALITY=	EMBEDDED_FOONOTES=
ENBED_FOOTNOTES_ONCE=	EMBEDDED_TITLES=
EMBED_TITLES_ONCE=	FITTOPAGE=
FORMULAS=	FROZEN_HEADERS=
FROZEN_ROWHEADERS=	GRIDLINES=
HIDDEN_COLUMNS=	HIDDEN_ROWS=
INDEX=	MSG_LEVEL=
ORIENTATION=	PAGE_ORDER_ACROSS=
PAGES_FITHEIGHT=	PAGES_FITWIDTH=
PRINT_AREA=	PRINT_FOOTER=
PRINT_FOOTER_MARGIN=	PRINT_HEADER=
PRINT_HEADER_MARGIN=	ROWBREAKS_COUNT=
ROWBREAKS_INTERVAL=	ROWCOLHEADINGS=
ROW_HEIGHTS=	ROW_REPEAT=
SCALE=	SHEET_INTERVAL=
SHEET_LABEL=	SHEET_NAME=
START_AT=	SUPPRESS_BYLINES=
TAB_COLOR=	TITLE_FOOTNOTE_NOBREAK=
TITLE_FOOTNOTE_WIDTH=	ZOOM=

These optional arguments will be explained in later chapters.

General Chapter Descriptions

This book is a book of examples, and as mentioned before, the examples are grouped into features and options that I feel have common traits. Each of the following chapters contain the following sections:

- A list of options to be discussed
- The syntax used to execute the options
- A description of the output generated
- The SAS code to produce the Excel workbook
- The output worksheets generated by the SAS code
- A conclusion

Chapter 3: ODS Excel Destination Actions

Introduction

The ODS Excel destination "Actions" are ODS commands dealing with the structure and content of the output Excel workbook. There are only a few commands classified as "Actions" for the ODS EXCEL statement, and they are "Close", "Exclude", "Select", and "Show". In this chapter I explain these actions and provide examples. First, after describing the actions, I want to demonstrate the simplest implementation that will enable you to create an Excel workbook.

The ODS TRACE statement is not actually part of the ODS Excel destination. However, it will help you identify the names of output ODS objects that you might want to SELECT or EXCLUDE when outputting data to your Excel files.

Table 3-1 – ODS TRACE Statement

Action Parameter	Description
ODS TRACE	Creates a listing in the SAS log that identifies ODS output objects generated by your code that can be controlled with the ODS EXCEL "SELECT" and "EXCLUDE" actions.

ODS Excel Destination Actions

In this chapter we will discuss the following SAS ODS Excel destination actions.

Table 3-2 – ODS Excel Destination Actions

Action Parameter	Options	Description
NONE	none	Sends the output Excel file to the SAS default output directory. Depending on your version of SAS the default directory is shown in the bottom left or right side of the display manager window.
EXCLUDE*	ALL, NONE, Exclusion(s)	Suppresses an ODS output object from the output Excel workbook.
SELECT*	ALL, NONE, Exclusion(s)	Includes an ODS output object in the output Excel workbook.
SHOW*	none	Writes to the SAS log the list of currently excluded or selected ODS output objects.
CLOSE	none	This action closes the destination. Any files currently associated with the ODS EXCEL statement are released by SAS and can now be opened by running Excel.

* The ODS Excel destination must be open to use this action.

A Simple ODS Excel Example and the CLOSE Action

The SAS documentation describes the simplest ODS EXCEL statement implementation, as shown below, with no arguments followed by some SAS procedural code and the ODS EXCEL statement with the CLOSE action argument. In this first example I am showing all of the screens that are generated by SAS 9.4 (TS1M3) using PC SAS on Windows. Future examples will only show the code and Excel output workbook or worksheets generated. In this example, I changed the SAS default output directory to "C:\Users\wmebe_000\Book_2016" by selecting the "Tools>Options>Change Current Folder>Change Folder" screen on the SAS toolbar. The folder must exist for the change to be effective. Typically, I would use the FILE= option, which will be explained in a later chapter, but I just wanted to show that it is not required.

A Simple ODS Excel Output Step

The code in SAS Code 3-1 is so basic that it does not name either the workbook or the worksheet that is generated by the code. This code accepts all of the default values provided by SAS. The output location is the current SAS directory, the workbook name is "sasexcl.xlsx", and the sheet name is "Print 1 - Data set SASHELP.S," which is the first 28 characters of the SAS Procedure and the SAS data set name.

SAS Code 3-1 – A Simple ODS Excel Output

```
ODS EXCEL;
PROC PRINT DATA=sashelp.shoes;
RUN;
ODS EXCEL CLOSE;
```

Figure 3-1 shows the SAS code, SAS log, and SAS default location of the output Excel Workbook. The LOG window also shows a NOTE that says "NOTE: Writing HTML Body file: sashtml.htm. This occurs because the ODS HTML location was not closed before running the SAS Code 3-1.

Figure 3-1 – SAS Code, Log, and Default Output Directory

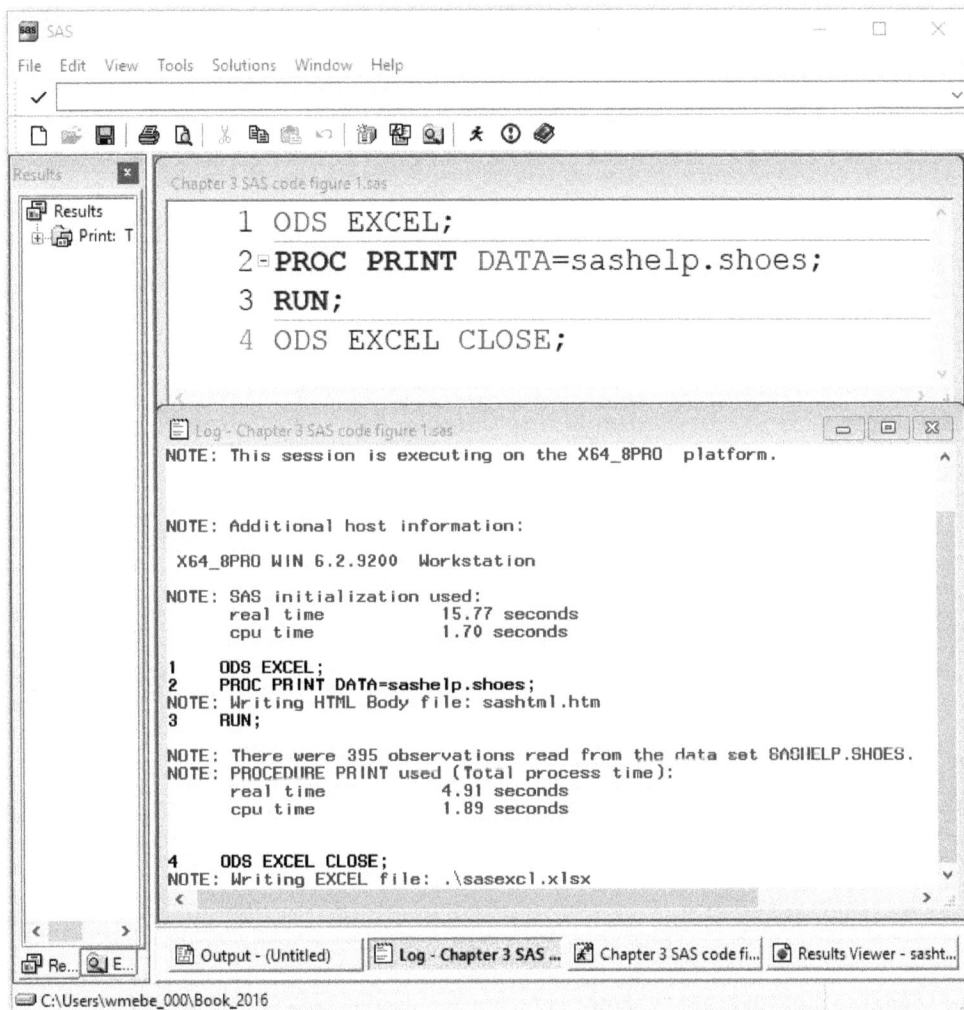

On the bottom of the PC SAS display manager window the default output directory name is listed. The actual location on the screen varies depending on the version of SAS that you are using. This form of execution selects the filename at execution time, while the FILE= option enables you to select an output filename. In this example, the name defaults to sasexcl.xlsx, but on other operating systems the default name might be different. Depending on the operating system that the SAS code was running on, and the TOOLS> Options> Preferences "Results" tab selections, the output EXCEL workbook can be forced open using EXCEL. This is shown in Figure 3-2.

Figure 3-2 – Excel Output from a Simple ODS EXCEL Statement

The windows output directory is shown in Figure 3-3; the directory was empty before the SAS code was executed. Afterward, the directory shows the output Excel workbook. Since the workbook was opened the temporary file generated by Excel is also visible.

Figure 3-3 – SAS Default Output Windows Directory

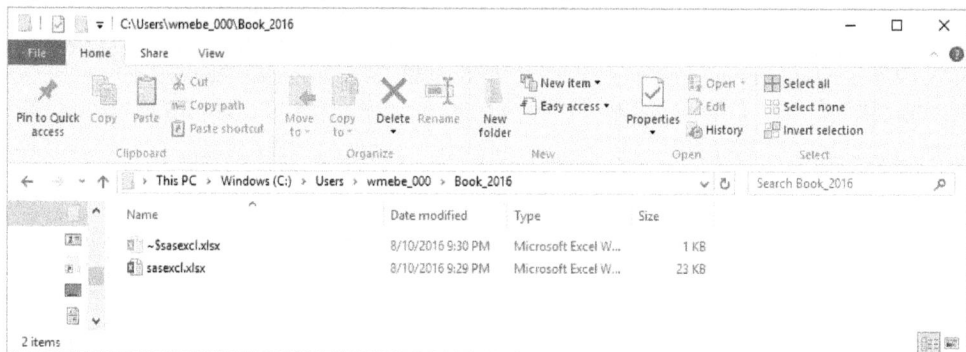

As you can see the simplest form of writing an Excel file with the ODS Excel destination is uncomplicated. When Microsoft released Version 2007 of Microsoft Office software, they changed the format of their Excel workbooks from a binary formatted file to an Open XML set of files stored in a compressed file structure. Most people, myself included, would call this a "ZIP" file. While this may not be what SAS or Microsoft call the file structure, if you change the ".xlsx" of the filename to ".zip" you can use the Winzip program to open the Excel workbook to see the underlying data elements. Figure 3-4 shows the internal directories of the Excel *.xlsx workbook. While I do not intend to expose all of the mysteries of the internal workings of an Excel workbook, I will say that the data is in there somewhere. Now you know where to look.

Figure 3-4 – Excel Output File XML Contents

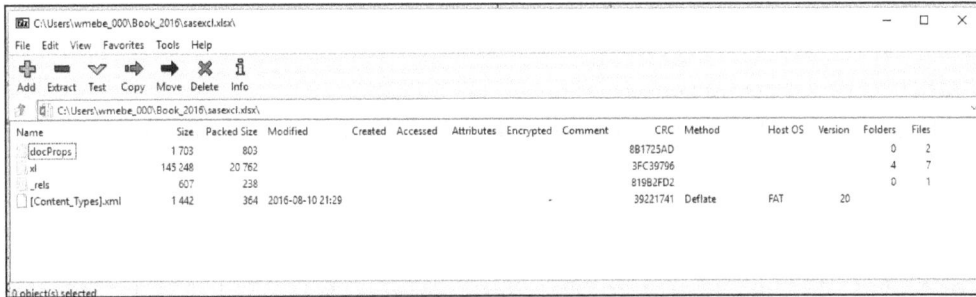

The first section shows how to create an Excel workbook from original data in one worksheet with all of the data variables and a new column indicating the SAS observation number. This is straightforward and generated an Excel workbook in native format for the *.xlsx output files.

Finding ODS Object Names Using the ODS TRACE Statement

The next step in our process is to find a way to identify the ODS output elements that are available to be output by ODS. In other words, if you do not know what you have, how do you know what you want to keep? We will use the ODS TRACE statement to identify output objects that can be output. Then I will demonstrate how to use the ODS EXCEL actions SHOW, SELECT, and EXCLUDE to finalize your output. By knowing how to show, select, or exclude output objects, you can be selective in what your output workbooks contain. The SAS Output Delivery System does a lot of work. The ODS TRACE statement is one feature of ODS that lets you look under the covers to identify internal objects used by the system to generate output. Information about each object generated for ODS by procedures and the DATA step is exposed in the SAS Log. This information can then be used at project development time to finalize this selection of output data objects. The simple commands ODS TRACE ON and ODS TRACE OFF turn this feature on and off. The code in SAS Code 3-2 shows how to use the ODS TRACE statement. Here it is turned on for only the UNIVARIATE procedure. When a computer can output something about everything it does, you can usually get more than you bargained for. In this example, I just wanted to see the output for one procedure.

Using the ODS TRACE Statement to Show PROC UNIVARIATE Objects

While I am showing the ODS TRACE, ODS EXCLUDE, ODS SELECT, and ODS SHOW statements as part of the ODS Excel destination, they are in fact separate ODS statements in their own right. These four ODS statements enable you to identify, choose, and verify the ODS objects that you want to output. They are described in this chapter because they work in conjunction with the ODS Excel Destination.

SAS Code 3-2 – A PROC UNIVARIATE Call Showing SAS ODS Output Objects

```
ODS TRACE ON;
PROC UNIVARIATE  DATA=sashelp.shoes
   (keep =  region stores product
    where=(region="Asia" and product="Boot"));
BY region;
variable stores;
freq stores;
RUN;
ODS TRACE OFF;
```

Log Output for SAS Code 3-2

This log listing shows the ODS elements output in the order that they were generated by PROC UNIVARIATE. Each "Output Added" notation in the log describes a new ODS output element created by the UNIVARIATE procedure. The "Name" of the ODS object can be used to either SELECT or EXCLUDE that part of the ODS output stream into the Excel workbook.

Log Output 3.1 – Log Listing of the ODS TRACE Commands

```
NOTE: This session is executing on the X64_8PRO  platform.

NOTE: Additional host information:

 X64_8PRO WIN 6.2.9200  Workstation

NOTE: SAS initialization used:
      real time             1.36 seconds
      cpu time              0.98 seconds

1     ODS TRACE ON;
2     PROC UNIVARIATE  DATA=sashelp.shoes
NOTE: Writing HTML Body file: sashtml.htm
3        (keep =  region stores product
4         where=(region="Asia" and product="Boot"));
5     BY region;
6     variable stores;
7     freq stores;
8     RUN;

Output Added:
-------------
Name:       Moments
Label:      Moments
Template:   base.univariate.Moments
Path:       Univariate.ByGroup1.Stores.Moments
-------------

Output Added:
-------------
Name:       BasicMeasures
Label:      Basic Measures of Location and Variability
```

```
Template:     base.univariate.Measures
Path:         Univariate.ByGroup1.Stores.BasicMeasures
------------

Output Added:
------------
Name:         TestsForLocation
Label:        Tests For Location
Template:     base.univariate.Location
Path:         Univariate.ByGroup1.Stores.TestsForLocation
------------

Output Added:
------------
Name:         Quantiles
Label:        Quantiles
Template:     base.univariate.Quantiles
Path:         Univariate.ByGroup1.Stores.Quantiles
------------

Output Added:
------------
Name:         ExtremeObs
Label:        Extreme Observations
Template:     base.univariate.ExtObs
Path:         Univariate.ByGroup1.Stores.ExtremeObs
------------
NOTE: The above message was for the following BY group:
      Region=Asia
NOTE: PROCEDURE UNIVARIATE used (Total process time):
      real time            0.75 seconds
      cpu time             0.51 seconds

9     ODS TRACE OFF;
```

SAS Output for SAS Code 3-2

Figure 3-5 shows the HTML output generated by PROC UNIVARIATE where Region equals "Asia", Product equals "Boot", and Variable equals "Stores" from the SASHELP.SHOES data set.

Figure 3-5 – PROC UNIVARIATE Output

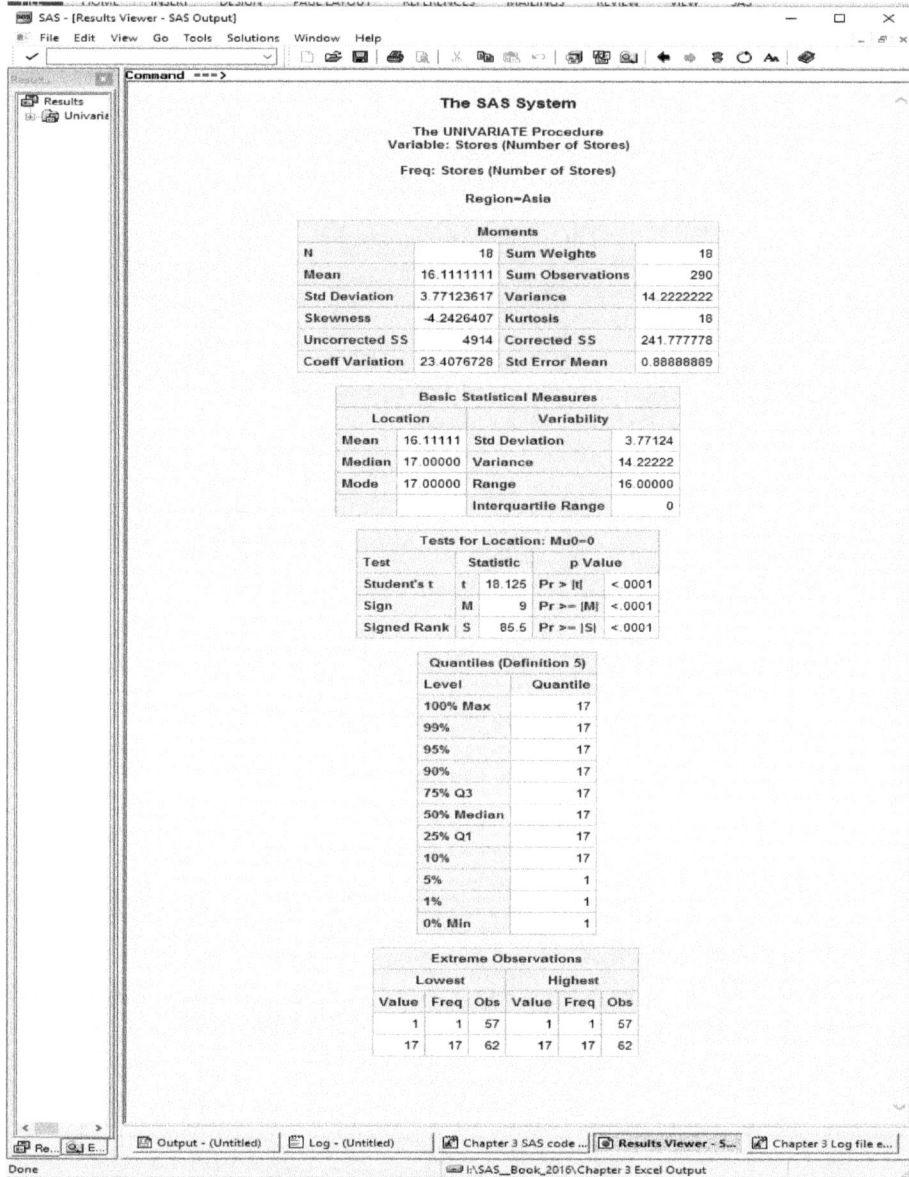

Each table of the output in the Results window is one of the ODS output objects generated by PROC UNIVARIATE. The tables are shown in the SAS log in the order in which they are generated. Two of the ODS output objects, "Moments" and "Quantiles" that were identified in SAS Code 3-2, will be used in both examples.

ODS EXCLUDE Statement

The ODS EXCUDE statement is a stand-alone ODS statement. It is discussed here because the syntax includes the name of an open and active ODS statement, when the ODS EXCLUDE statement is encountered in the SAS code. The purpose of the ODS EXCLUDE statement is to prevent ODS objects from being output. The ODS TRACE statement shown in SAS Code 3-2 enables you to find the names of the ODS objects that are available for output. SAS Code 3-3 uses the ODS EXCLUDE statement to prevent the "Moments" and "Quantiles" ODS objects from being output to the Excel workbook, but all of the other objects from the UNIVARIATE procedure are output. I am also using an ODS Excel suboption that will be described later so that I can show all of the output on one page of an Excel workbook. When SAS Code 3-3 is executed, the SAS Results window will look like Figure 3-5 because nothing was "EXCLUDED" from that ODS output stream. However, the Excel output will not include the "Moments" and "Quantiles" ODS objects and is shown in Figure 3-6.

SAS Code 3-3 – Excluding Moments and Quantiles from the Excel Output Workbook

```
ODS EXCEL options(Sheet_interval='NONE');
ODS EXCEL EXCLUDE Moments Quantiles;
PROC UNIVARIATE  DATA=sashelp.shoes
   (keep =  region stores product
    where=(region="Asia" and product="Boot"));
variable stores;
freq stores;
RUN;
ODS EXCEL CLOSE;
```

Figure 3-6 – Excel Output Workbook Excluding Moments and Quantiles

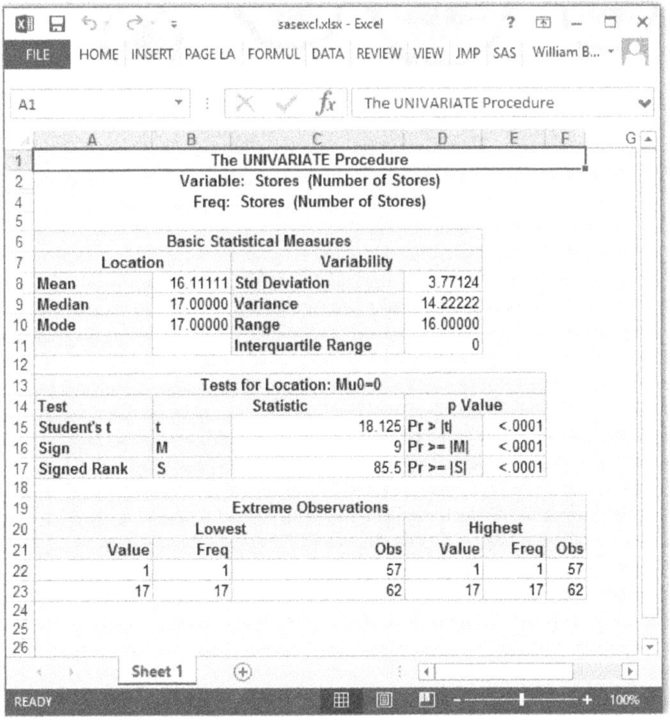

The UNIVARIATE Procedure
Variable: Stores (Number of Stores)
Freq: Stores (Number of Stores)

Basic Statistical Measures			
Location		Variability	
Mean	16.11111	Std Deviation	3.77124
Median	17.00000	Variance	14.22222
Mode	17.00000	Range	16.00000
		Interquartile Range	0

Tests for Location: Mu0=0				
Test	Statistic		p Value	
Student's t	t	18.125	Pr > |t|	<.0001
Sign	M	9	Pr >= |M|	<.0001
Signed Rank	S	85.5	Pr >= |S|	<.0001

Extreme Observations					
Lowest			Highest		
Value	Freq	Obs	Value	Freq	Obs
1	1	57	1	1	57
17	17	62	17	17	62

ODS SELECT Statement

The ODS SELECT statement is a stand-alone ODS statement. It is discussed here because the syntax includes the name of an ODS statement that is open and active when the ODS SELECT statement is encountered within the SAS code. The purpose of the ODS SELECT statement is to include ODS objects in the output. The ODS TRACE statement shown in SAS Code 3-2 enables you to find the names of the ODS objects that are available for output. The code in SAS Code 3-4 uses the ODS SELECT statement to include the "Moments" and "Quantiles" ODS objects in the output to the Excel workbook, but all of the other objects from the UNIVARIATE procedure are suppressed from the output. I am also using an ODS EXCEL suboption that will be described later so that I can show all of the output on one page of an Excel workbook. When SAS Code 3-4 is executed, the SAS Results window will look like Figure 3-5 because nothing was "EXCLUDED" from that ODS output stream. However, the Excel output will only include the "Moments" and "Quantiles" ODS objects and is shown in Figure 3-7.

SAS Code 3-4 - SAS Code to Select Moments and Quantiles in the Excel Output Workbook

```
ODS EXCEL options(Sheet_interval='NONE');
ODS EXCEL SELECT Moments Quantiles;
PROC UNIVARIATE  DATA=sashelp.shoes
    (keep =  region stores product
     where=(region="Asia" and product="Boot"));
variable stores;
freq stores;
RUN;
ODS EXCEL CLOSE;
```

Figure 3-7 - Excel Output Workbook Showing Moments and Quantiles

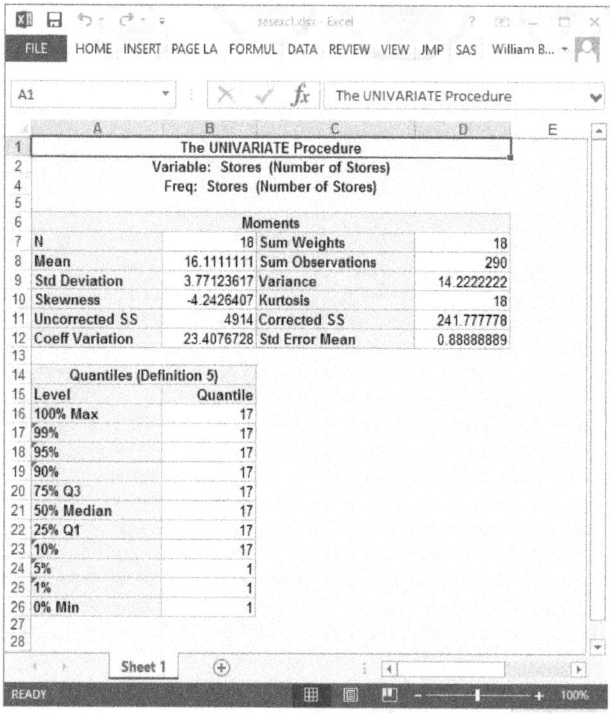

The UNIVARIATE Procedure
Variable: Stores (Number of Stores)
Freq: Stores (Number of Stores)

Moments			
N	18	Sum Weights	18
Mean	16.1111111	Sum Observations	290
Std Deviation	3.77123617	Variance	14.2222222
Skewness	-4.2426407	Kurtosis	18
Uncorrected SS	4914	Corrected SS	241.777778
Coeff Variation	23.4076728	Std Error Mean	0.88888889

Quantiles (Definition 5)	
Level	Quantile
100% Max	17
99%	17
95%	17
90%	17
75% Q3	17
50% Median	17
25% Q1	17
10%	17
5%	1
1%	1
0% Min	1

ODS SHOW Statement

The ODS SHOW statement is a stand-alone ODS statement. It is discussed here because the syntax includes the name of an ODS statement that is open and active when the ODS SHOW statement is encountered within the SAS code. The purpose of the ODS SHOW statement is to indicate the status of ODS objects in the output. The ODS TRACE statement shown in SAS Code 3-2 enables you to find the names of the ODS objects that are available for output. SAS Code 3-5 uses the ODS SELECT statement to include the "Moments" and "Quantiles" ODS objects in the output to the Excel workbook, but all of the other objects from the UNIVARIATE procedure are suppressed from the output. I am also using an ODS EXCEL suboption that will be described later so that I can show all of the output on one page of an Excel workbook. When SAS Code 3-5 is executed, the SAS Results window will look like Figure 3-5 because nothing was "EXCLUDED" from that ODS output stream. However, the Excel output will only include the "Moments" and "Quantiles" ODS objects and is the same as shown in Figure 3-7. I have also included an ODS EXCEL SHOW statement; the results of this statement are shown in SAS Log 3-2 below.

SAS Code 3-5 – SAS Code to Select Moments and Quantiles in the Excel Output Workbook

```
ODS EXCEL options(Sheet_interval='NONE');
ODS EXCEL SELECT Moments Quantiles;
ODS EXCEL SHOW;
PROC UNIVARIATE   DATA=sashelp.shoes
   (keep =  region stores product
    where=(region="Asia" and product="Boot"));
variable stores;
freq stores;
RUN;
ODS EXCEL CLOSE;
```

Log Output 3-2 – Example of the Output of the ODS SHOW Statement

```
65    ODS EXCEL options(Sheet_interval='NONE');
66    ODS EXCEL SELECT Moments Quantiles;
67    ODS EXCEL SHOW;
Current EXCEL select list is:
1. Moments
2. Quantiles
68    PROC UNIVARIATE   DATA=sashelp.shoes
69       (keep =  region stores product
70        where=(region="Asia" and product="Boot"));
71    variable stores;
72    freq stores;
73    RUN;

NOTE: PROCEDURE UNIVARIATE used (Total process time):
      real time            0.18 seconds
      cpu time             0.11 seconds


74    ODS EXCEL CLOSE;
NOTE: Writing EXCEL file: .\sasexcl.xlsx
```

Conclusion

The number of ODS EXCEL actions are limited but they impact the output workbook and the log. The ODS SELECT and ODS EXCLUDE statements produced the same Results window entries because the commands were not issued for the standard output HTML ODS output stream. However, the Excel workbook that was generated was different for each command. Log entries for the ODS SHOW output were shown in Log Output 3.2 and they produced very different Excel files. These ODS EXCEL actions enable you to selectively output elements of the ODS generated output data. This gives you full control over the output that is generated and what you want to include in an Excel workbook.

Chapter 4: Setting Excel Document Property Values

Introduction

This chapter addresses ODS EXCEL destination options that enable you to enter text into the "Excel Properties Sheet" directly from SAS. This is a time-saving option that enables you to update properties at the same time that your workbook is created. This also eliminates the need to explain why your workbook was modified after it was created. All of the options described in this chapter accept a text string as input.

Options to Modify the Excel Document Properties Sheet

In this chapter we will discuss the following ODS EXCEL destination options that address ways to apply text strings to the items in the Excel document properties sheet. The options listed in Table 4-1 identify the ways that SAS can send text strings to the Excel Document Properties Sheet.

Table 4-1 – ODS Excel Destination Options to Modify the Excel Document Property Sheet Values

Action Parameter	Description
AUTHOR	adds an author information string to the Excel document properties.
CATEGORY	adds a category information string to the Excel document properties.
COMMENTS	adds a comment string to the Excel document properties.
FILE	selects a path and filename for your output Excel workbook.
KEYWORDS	adds a keyword string to the Excel document properties.
STATUS	creates a status. You can provide a text string that can be seen in the Excel document properties.

Action Parameter	Description
SASDATE	saves the SAS date in the Excel property sheet instead of the default Excel date.
TITLE	adds a title string to the Excel document properties.

Default Excel Property Sheets

The SAS documentation describes the simplest ODS EXCEL statement implementation as shown below in SAS Code 4-1. Writing the ODS EXCEL statement code with no arguments (also called Options) invokes output to the default home directory. SAS Code 4-1 shows the best practice of using the FILE= option to identify an output Excel workbook. When using PC SAS, the current folder can be identified in two ways. Either by looking at the bottom of the SAS Display Manager window (bottom left or bottom right depending on your SAS version) or by selecting the "Change Folder" screen of the "Tools>Options>Change Current Folder" menu selection.

The ODS EXCEL statement with the CLOSE action argument releases the Excel workbook for use by other activities, including opening the workbook with Excel or some other program.

SAS Code 4-1 – Code That Does Not Change the Default Excel Document Property Values

```
*Code that does not Change Default Excel document property values;

ODS EXCEL;
PROC PRINT DATA=SASHELP.Shoes;
RUN;
ODS EXCEL CLOSE;




*Code that does not Change Default Excel document property values;
*Using the FILE= option;

ODS EXCEL FILE="C:\Users\wmebe_000\Book_2016\Excel_workbook.xlsx";
PROC PRINT DATA=SASHELP.Shoes;
RUN;
ODS EXCEL CLOSE;
```

Figure 4-1 – Excel Document Property Sheet with Default Properties

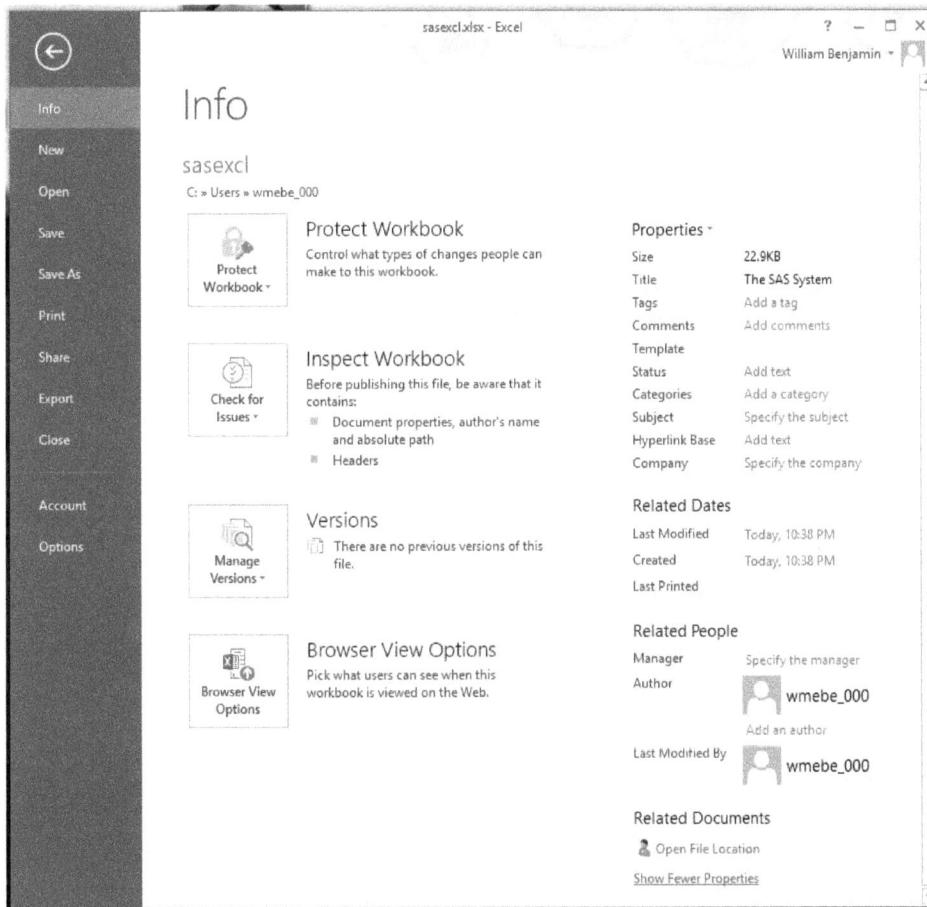

Note that the **Author** and **Last Modified** fields use field values that are assigned when the account login was created, and the Title value is the default SAS title text.

Modify Excel Property Sheets with the ODS EXCEL Destination

The next example adds character strings to the Excel document properties sheet. Each of the following options affect the Excel property sheet, and are shown in Figure 4-2.

- AUTHOR
- CATEGORY
- COMMENTS
- KEYWORDS
- STATUS
- TITLE

SAS Code 4-2 –Setting the Excel Document Property Values

```
ODS EXCEL
  AUTHOR    = 'AUTHOR    - text-string 1'
  CATEGORY  = 'CATEGORY  - text-string 2'
  COMMENTS  = 'COMMENTS  - text-string 3'
  KEYWORDS  = 'KEYWORDS  - text-string 4'
  STATUS    = 'STATUS    - text-string 5'
  TITLE     = 'TITLE     - text-string 6'
;
PROC PRINT DATA=SASHELP.Shoes;
RUN;
ODS EXCEL CLOSE;
```

Note that most SAS syntax diagrams do not show spaces between letters and symbols of the SAS instructions. This code demonstrates that you can add spaces to make your code more readable.

Figure 4-2 – Excel Document Property Sheet with SAS Generated Properties

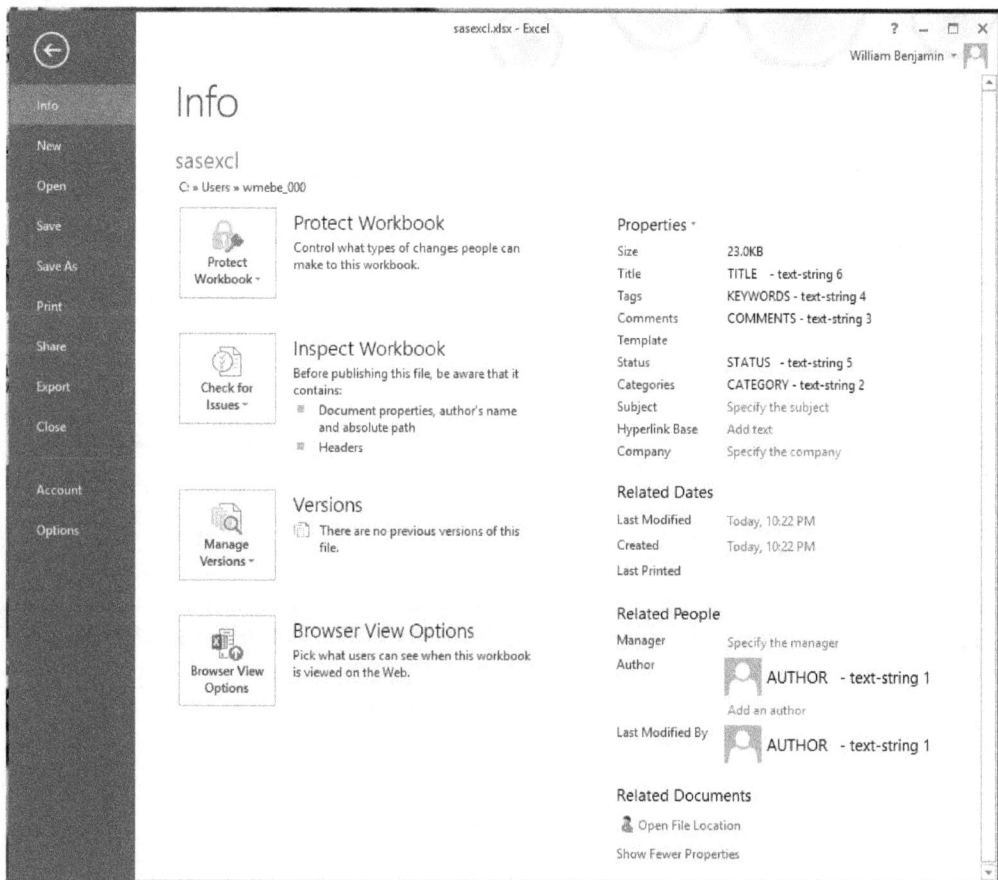

Notice that the text for the KEYWORDS option appears to the right of the **Tags** field on the properties sheet.

Excel Property Sheet Date/Time Fields without the SASDATE Option

SAS Code 4-3 – Excel Document Property Sheet without the SASDATE Option

```
ODS EXCEL;
PROC PRINT DATA=sashelp.shoes;
run;
ODS EXCEL CLOSE;
```

Figure 4-3 – Excel Document Property Sheet without the SASDATE Option

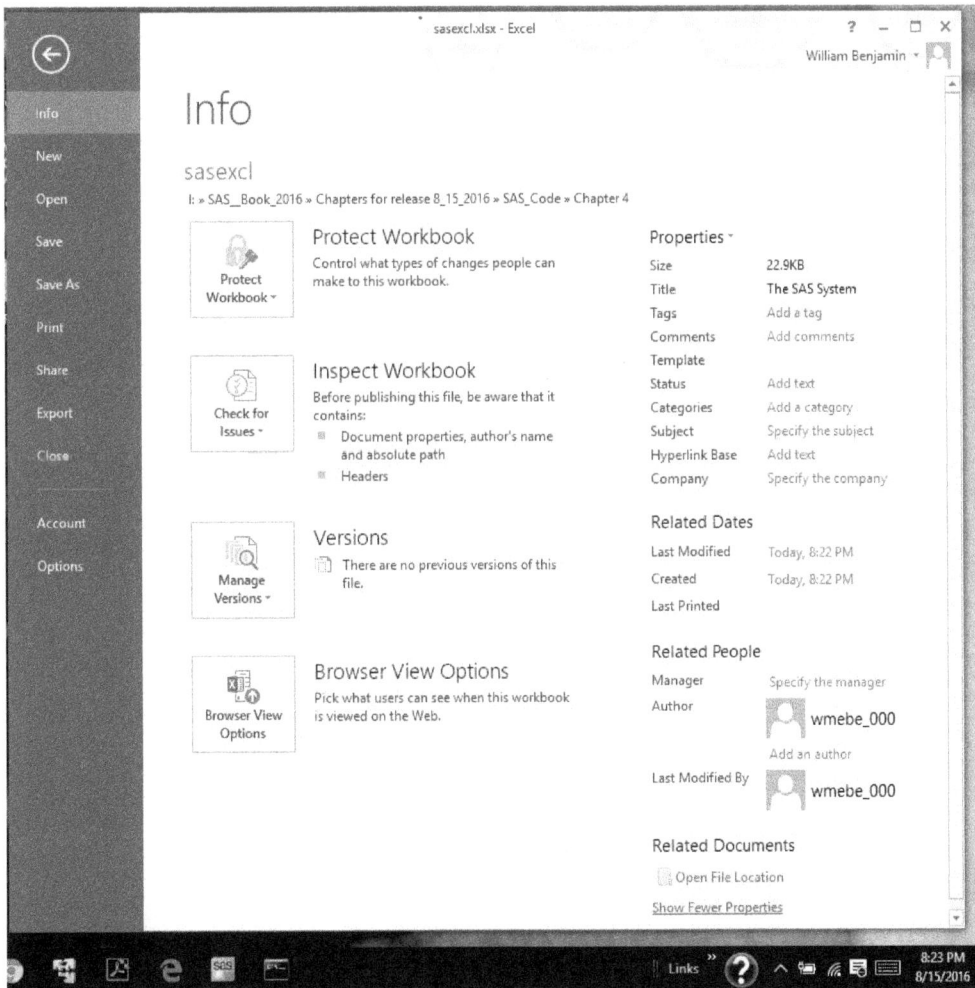

Excel Property Sheet Date/Time Fields with the SASDATE Option

The SAS Code 4-4 example with the results shown in Figure 4-4 was produced using the SASDATE option. The code executed in SAS Code 4-3 was executed before the code in SAS Code 4-4. But the dates in Figure 4-4 are before the dates shown in Figure 4-3. This is because the

SASDATE option uses the DATE/TIME from when the SAS session started as the value entered for the dates.

SAS Code 4-4 – Excel Document Property Sheet with the SASDATE Option

```
ODS EXCEL SASDATE;
PROC PRINT DATA=sashelp.shoes;
run;
ODS EXCEL CLOSE;
```

Figure 4-4 – Excel Document Property Sheet with the SASDATE Option

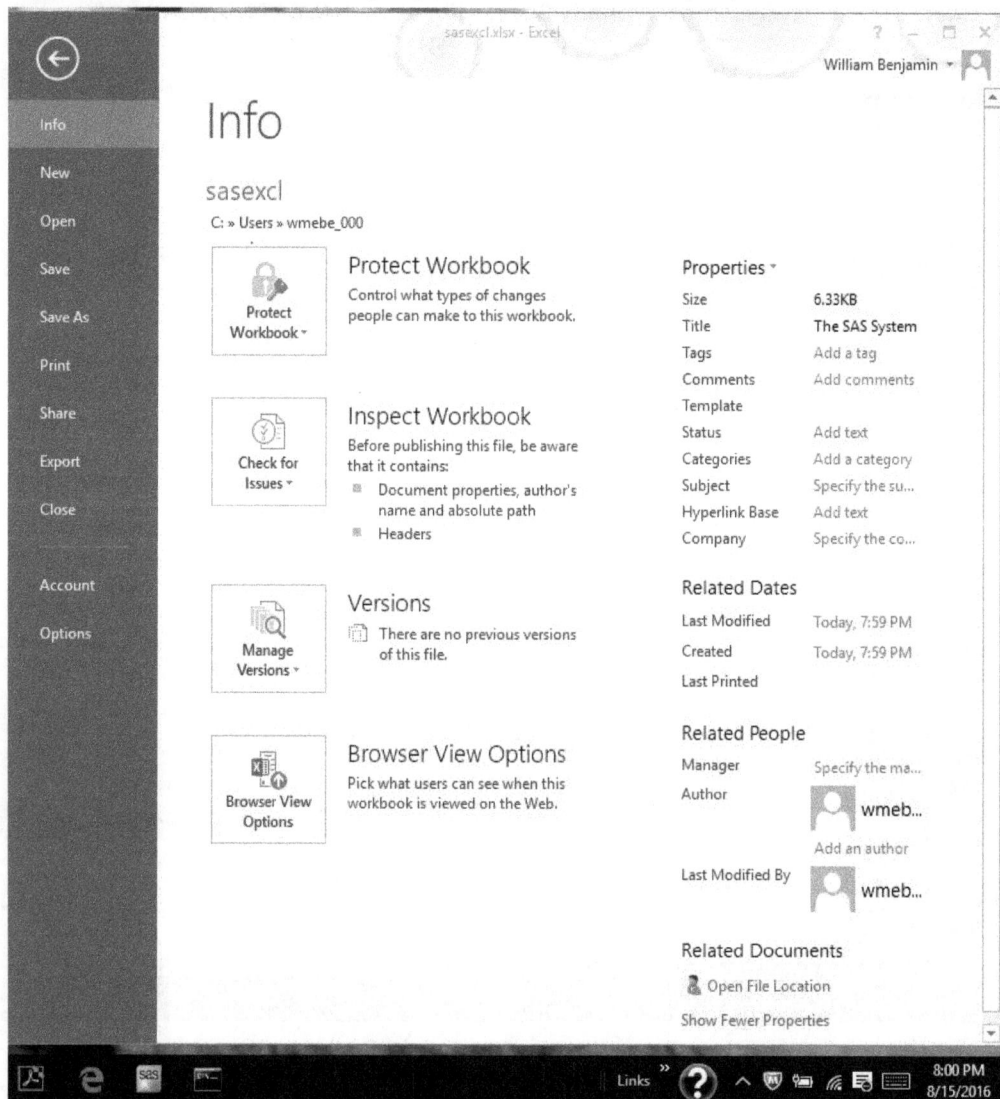

Conclusion

A number of ODS EXCEL options enable you to modify the Excel document property sheet. You can populate the values directly from your SAS program. The strings that you specify can also be SAS macro variables. This gives you full control over the data generated and what you want to output to the Excel workbook property sheet fields. Many times these values are manually updated after the workbook is generated. If your program writes only one workbook, this might not be an issue for you. However, if your SAS code produced several or hundreds of workbooks, you might not want to take the time to update each one manually.

Chapter 5: Options That Affect the Workbook

Introduction

This chapter will discuss the ODS EXCEL ID= option and some suboptions that affect output at the level of the whole workbook. In Chapter 2, the SASHELP.SHOES data was filtered with the WHERE clause to only include data from the "ASIA" region. This was done to enable display of the data on one screen image. Unless otherwise noted, all of the examples in this and the following chapters will be using the following data set called "ASIA_ONLY" that was shown in Chapter 2.

Options and suboptions discussed in this chapter affect creating multiple output worksheets placing blank sheets into a workbook, setting up a Table of Contents or Index for the workbook with hyperlinks, suppressing messages, changing the tab colors, and messages and zoom level of the worksheets. Most programmers or people who work with Excel workbooks take these actions after the file is already created. This chapter shows you how to provide these features before you give the workbook to someone for review. These features can be a real time saver.

ODS Excel Destination Actions

In this chapter we will discuss the following SAS ODS Excel destination topics. If the SAS code shows the topic within the syntax "OPTIONS= (topic)", then the topic is a "SUBOPTION" of the SAS ODS Excel destination.

Table 5-1–ODS Excel Destination Actions

Action Parameter	Options	Description
FILE	'file-specification'	This option defines the output file location of the Excel workbook to be created. This can be any valid file specification for the system that you are using. This ODS destination only creates files formatted as *.xlsx Excel workbooks.
ID	'numeric-positive-integer', or a series of characters that begin with a letter or underscore that can contain letters, numbers, and underscores	This option allows the creation of multiple instances of the same destination at the same time, each can have different options. The identifier will specify another instance of a destination that is already open, and the ID= option must directly follow the destination name. It is also possible to omit the ID= by using a name or number to identify the instance.
BLANK_SHEET	'string', Default = NONE	A blank worksheet is created in the workbook and the value of the 'string' is used as the name of the worksheet. Excel permits a tab label to be 32 characters, but the string that you provide cannot be longer than 28 characters because SAS reserves 4 characters to ensure any generated tab names are unique. The name must have a length of at least one character. The four characters are reserved at the end of the sheet name as a worksheet counter used to create unique worksheet names by adding a numeric counter value as a suffix to the string value.
CONTENTS	'OFF', 'ON', 'YES', 'NO', Default = 'OFF'	Positive responses to this option create a FIRST worksheet that contains a table of contents for the workbook.
INDEX	'OFF', 'ON', 'YES', 'NO', Default = 'OFF'	Positive responses to this option create a FINAL/LAST worksheet that contains an index of all worksheets for the workbook.
MSG_LEVEL	'string', Default = 'NO'	Messages from Excel are suppressed.
TAB_COLOR	'string'	The string supplied identifies the color of the tab for the next worksheet. You can use the following methods to describe the tab color. For the 'RGB' and 'RGBA' color definitions the percent signs are optional. A string that names a valid Excel color pallet name, like 'BLUE'. A string of 6 hexadecimal digits that describe a color definition, like '#0000ff'. A string that describes a RED, GREEN, BLUE set of colors, like 'RGB(0,0,100)' A string that describes a RED, GREEN, BLUE set of colors with a transparency index, like

Action Parameter	Options	Description
		'RGBA(0,0,100%,0.5)' *
ZOOM	'number', Default = 100	This setting describes the zoom level of the worksheet.

* http://ebanshi.cc/questions/3207084/can-i-set-a-rgba-color-within-a-color-theme

The FILE= Option

The FILE= option defines the output file location of the Excel workbook to be created. This can be any valid file specification for the system that you are using. The ODS destination only creates files formatted as *.xlsx Excel workbooks.

SAS Code 5-1 – Using the FILE= Option to Define an Output Excel Workbook

```
ods excel file="&path.\Excel_output_workbook.xlsx" ;
proc print data=asia_only;
run;
ods excel close;
```

Figure 5-1 – A Simple Excel File Defined by the FILE= Option of the ODS Excel Destination.

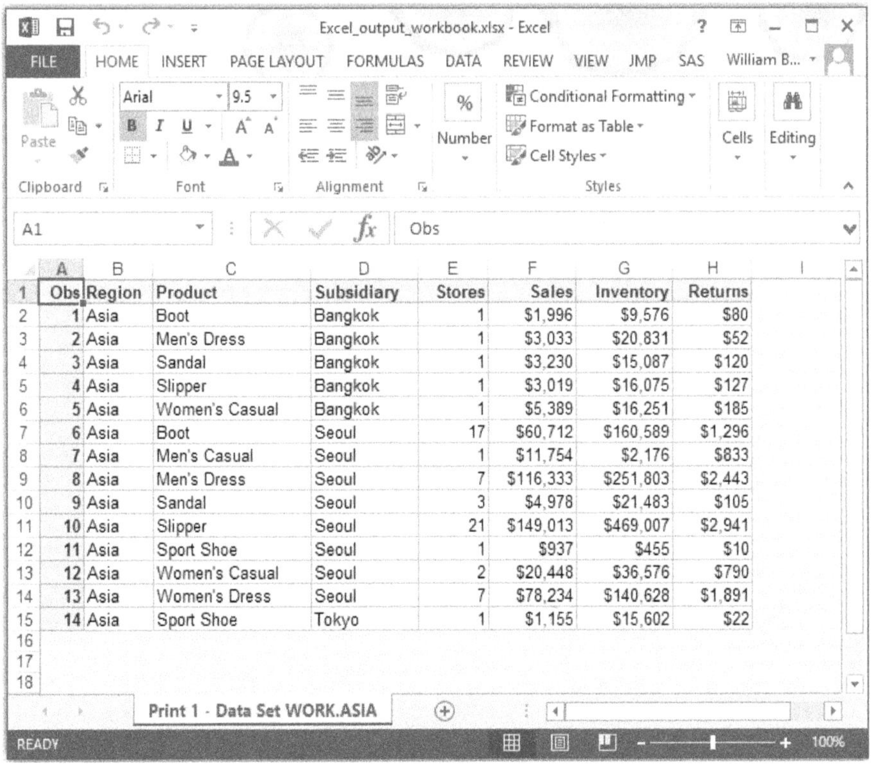

Obs	Region	Product	Subsidiary	Stores	Sales	Inventory	Returns
1	Asia	Boot	Bangkok	1	$1,996	$9,576	$80
2	Asia	Men's Dress	Bangkok	1	$3,033	$20,831	$52
3	Asia	Sandal	Bangkok	1	$3,230	$15,087	$120
4	Asia	Slipper	Bangkok	1	$3,019	$16,075	$127
5	Asia	Women's Casual	Bangkok	1	$5,389	$16,251	$185
6	Asia	Boot	Seoul	17	$60,712	$160,589	$1,296
7	Asia	Men's Casual	Seoul	1	$11,754	$2,176	$833
8	Asia	Men's Dress	Seoul	7	$116,333	$251,803	$2,443
9	Asia	Sandal	Seoul	3	$4,978	$21,483	$105
10	Asia	Slipper	Seoul	21	$149,013	$469,007	$2,941
11	Asia	Sport Shoe	Seoul	1	$937	$455	$10
12	Asia	Women's Casual	Seoul	2	$20,448	$36,576	$790
13	Asia	Women's Dress	Seoul	7	$78,234	$140,628	$1,891
14	Asia	Sport Shoe	Tokyo	1	$1,155	$15,602	$22

The ID= Option

The ID= option enables you to create multiple instances of ODS output from the same SAS code. Having different ID values enables you to apply different options and suboptions to separate output. The ID= option can be very useful if you need multiple copies of the same data. If you need a working Excel spreadsheet, a printable Excel workbook, or a PDF for outside delivery, you can use the ID= option to produce these copies in one process. The ability to do this enables you to have a working copy to markup and verify, a production copy to ship, and a PDF copy for documentation, without rerunning the job. My example here produces three files, two Excel workbooks and a PDF file. I will discuss these files separately. In the SAS Code 5-2 example I did not use the ID= option on the first ODS EXCEL statement to show that it is not required for the first instance, only for later ODS file definitions.

SAS Code 5-2 – Using the ODS ID= Option to Produce Multiple Outputs

```
ods excel file="&path.\ID_no_Style.xlsx" ;
ods excel (id=harvest) style=harvest
          file="&path.\ID_harvest_Style.xlsx" ;
ods pdf    (id=Sapphire) style=Sapphire
          file='ID_Sapphire_Style.pdf';

proc print data=asia_only;
run;

ods excel close;
ods excel (id=harvest)  close;
ods pdf    (id=Sapphire) close;
```

Figure 5-2 is an output Excel workbook. This workbook has no updates to the output workbook, it shows only the defaults, and this is what SAS will normally create.

Figure 5-2 – ODS Excel Output Using No Style or ID= Option (the defaults)

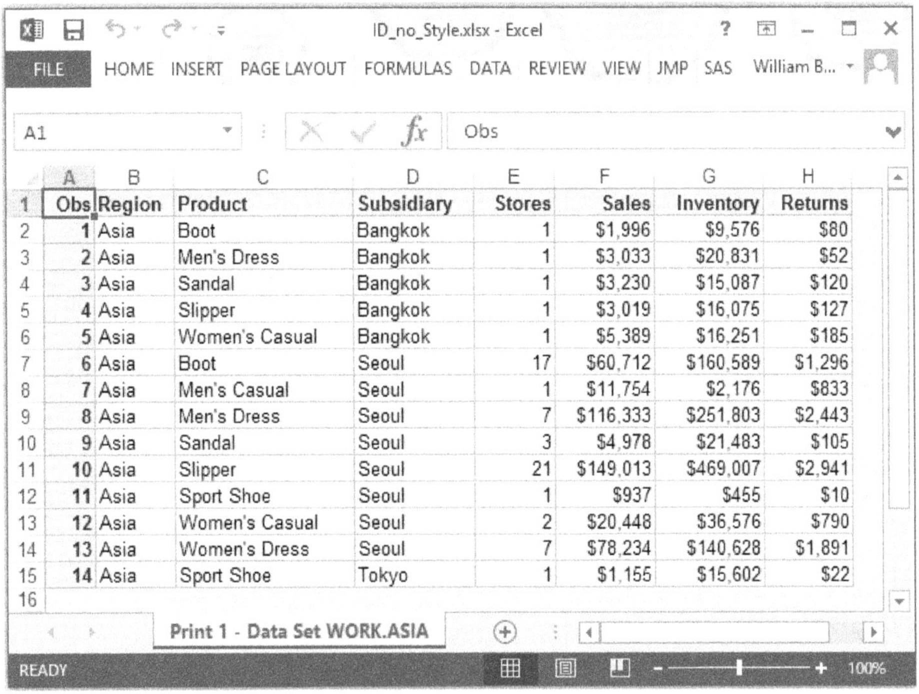

The output in the next image, Figure 5-3, was created using the SAS maintained STYLE called HARVEST. Notice the different colors in the row and column headers. The ID= option identified this file as HARVEST, during the opening and closing of the Excel Workbook. Any options assigned during this ODS FILE statement and sent to the output file at the close are unique to this Excel Workbook.

Figure 5-3 – ODS Excel Output Using Harvest Style and the ID= Option

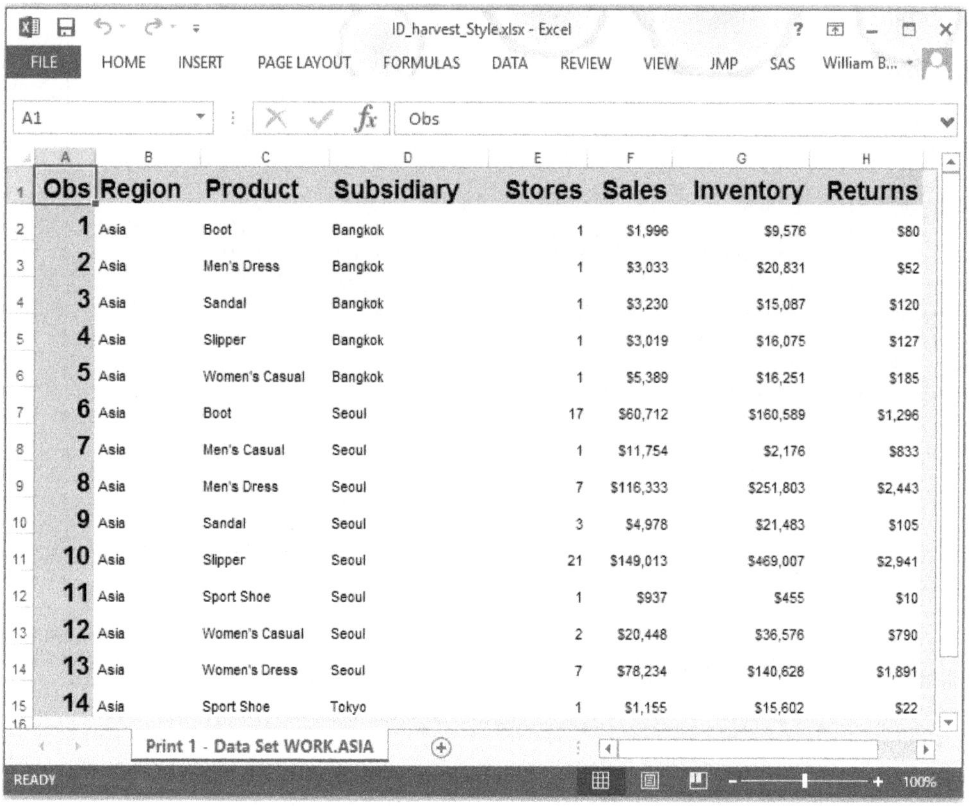

The next image, Figure 5-4, is the result of the ODS PDF statement. This output was generated with the SAS supported style called Sapphire. Once again, the ODS PDF statement defined the output options. In the ID= option where the ID was set to "SAPPHIRE", the output is a PDF file.

Figure 5-4 – ODS PDF Output Using the Sapphire Style and the ID= Option

The SAS System   1

Obs	Region	Product	Subsidiary	Stores	Sales	Inventory	Returns
1	Asia	Boot	Bangkok	1	$1,996	$9,576	$80
2	Asia	Men's Dress	Bangkok	1	$3,033	$20,831	$52
3	Asia	Sandal	Bangkok	1	$3,230	$15,087	$120
4	Asia	Slipper	Bangkok	1	$3,019	$16,075	$127
5	Asia	Women's Casual	Bangkok	1	$5,389	$16,251	$185
6	Asia	Boot	Seoul	17	$60,712	$160,589	$1,296
7	Asia	Men's Casual	Seoul	1	$11,754	$2,176	$833
8	Asia	Men's Dress	Seoul	7	$116,333	$251,803	$2,443
9	Asia	Sandal	Seoul	3	$4,978	$21,483	$105
10	Asia	Slipper	Seoul	21	$149,013	$469,007	$2,941
11	Asia	Sport Shoe	Seoul	1	$937	$455	$10
12	Asia	Women's Casual	Seoul	2	$20,448	$36,576	$790
13	Asia	Women's Dress	Seoul	7	$78,234	$140,628	$1,891
14	Asia	Sport Shoe	Tokyo	1	$1,155	$15,602	$22

The BLANK_SHEET= Option

This option produces an empty worksheet in the Excel workbook. But why would you want a blank worksheet? I use them for temporary scratch pads inside an Excel workbook. I often call VBS (Visual Basic Scripts) or VBA (Visual Basic for Applications) routines from my SAS code (see my *Exchanging Data Between SAS(R) and Microsoft Excel* [Cary, NC: SAS Institute, 2015], chaps. 12-14, for more information). These blank worksheets could be used by these windows or Excel routines and then deleted before closing the workbook. You might be able to think of other applications for creating a blank worksheet in an Excel workbook. The ODS command can be inserted into your SAS code at different places, and does not need to be the only option used. SAS Code 5-3 shows how to use this option to insert a blank sheet after producing an output sheet with data.

SAS Code 5-3 – Using the BLANK_SHEET= Option to Insert a Named Sheet into the Output Workbook

```
ods excel file="&path.\Empty_Workbook.xlsx";
proc print data=asia_only;
run;
ods excel options(BLANK_SHEET='Empty Sheet');
ods excel close;
```

Figure 5-5 – Excel Output of a Workbook with Data in One Sheet and a Blank Worksheet

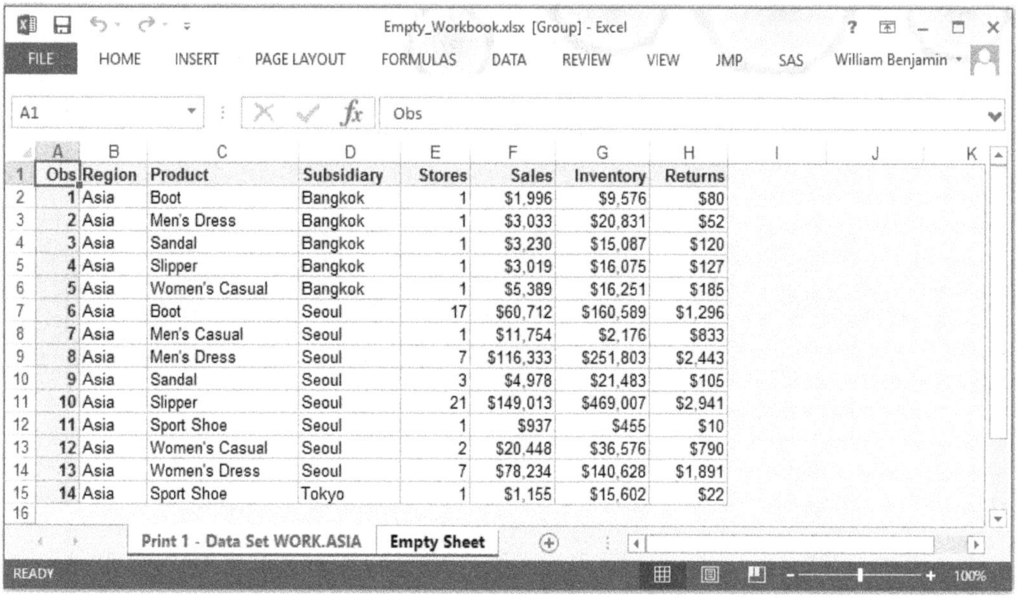

The CONTENTS= Option

The CONTENTS= option will build a Table of Contents listing for every worksheet in your workbook, and it will provide a hyperlink to enable you to jump directly to the sheet from the Table of Contents. This makes the job of listing the contents and enabling access to your data much easier. In Figure 5-6 below I am holding the cursor over cell A3 and the Excel options to select the cell or to follow the hyperlink are shown, but the cursor is not. Of course, your computer will show a different link path than mine. Here I am using the SASHELP.SHOES data set to show multiple pages. The CONTENTS= and INDEX= options produce similar results. The INDEX= option produces an abbreviated output.

SAS Code 5-4–Using the CONTENTS= Option to Produce Hyperlinks to Data Pages

```
ods excel file="&path.\Table_Of_Contents_Workbook.xlsx"
          options(CONTENTS='ON');
proc print data=sashelp.shoes;
by region;
run;
ods excel close;
```

Figure 5-6 – The Table of Contents Page of an Excel Workbook

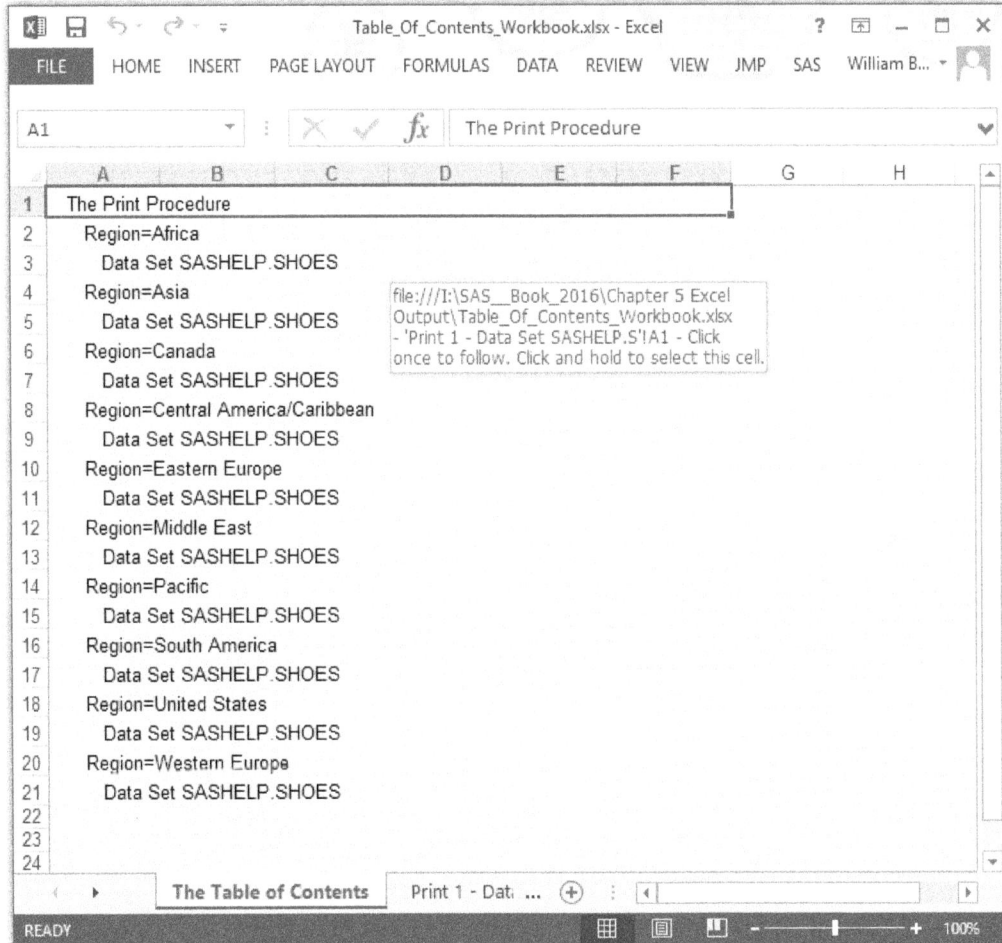

The INDEX= Option

The INDEX option will build a Table of Contents listing for every worksheet in your workbook, and it will provide a hyperlink to enable you to jump directly to the sheet from the Table of Contents. This makes the job of listing the contents and enabling access to your data much easier. In Figure 5-7 below I am holding the cursor over cell A1 and the Excel options to select the cell or to follow the hyperlink are shown, but the cursor is not. Of course, your computer will show a different link path than mine. Here I am using the SASHELP.SHOES data set to show multiple pages. The CONTENTS= and INDEX= options produce the similar results, The CONTENTS= option produces an expanded "Table of Contents" output.

SAS Code 5-5 – Using the INDEX= Option to Produce Hyperlinks to Data Pages

```
ods excel file="&path.\INDEX_Of_Workbook.xlsx"
          options(INDEX='ON');
proc print data=sashelp.shoes;
by region;
run;
ods excel close;
```

Figure 5-7 – The Index Page of an Excel Workbook

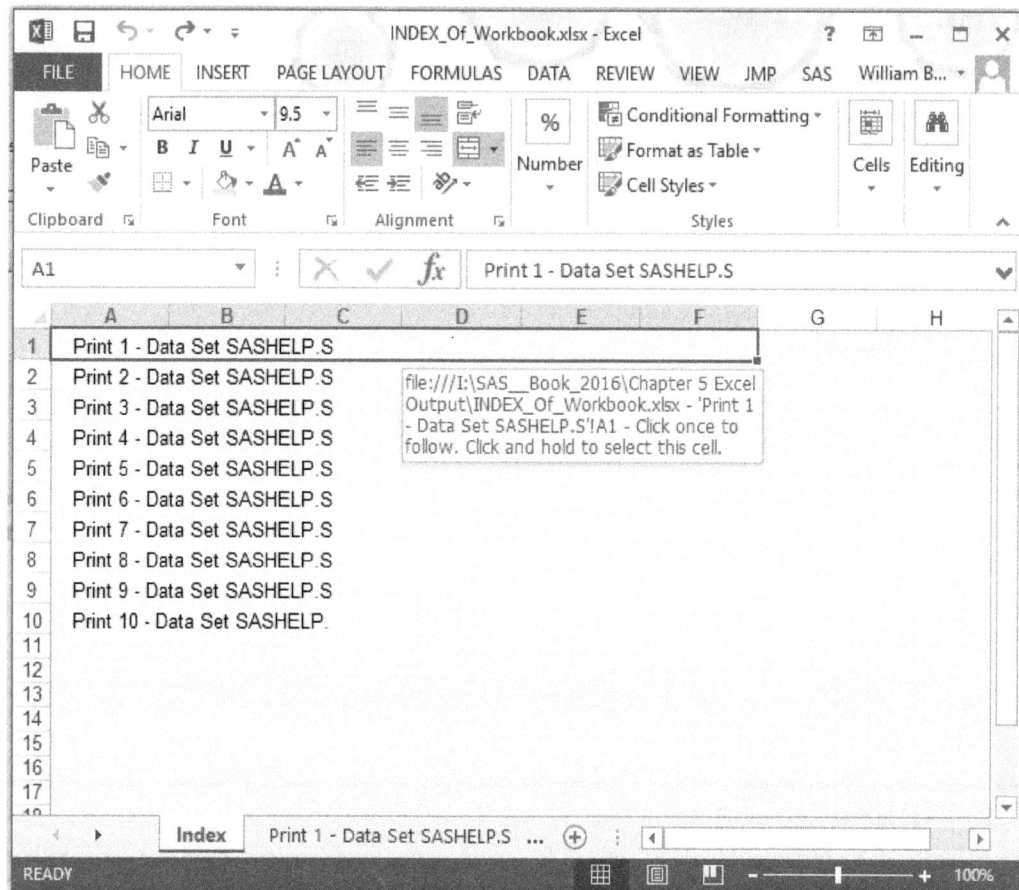

The MSG_LEVEL= Option

I had a lot of trouble with this suboption. After several emails with SAS Technical Support, I was told the option was to identify options from the EXCELXP Tagset that were deprecated in the ODS Excel destination. So I attempted to determine what type of responses I would get when using this option. SAS Code 5-6 shows four ways that I tried to run the code with the MSG_LEVEL= suboption. Well I set the value to 'NO', 'YES', 'any string', and did not even code the suboption. Each time I got similar result. Shown in the SAS Log 5-1 text box below.

SAS Code 5-6 – Using the MSG_LEVEL= Option

```
ods excel file="&path.\MSG_LEVEL_No_in_Workbook.xlsx"
          options(
          /* MSG_LEVEL='NO'                */
          /* MSG_LEVEL='YES'               */
          /* MSG_LEVEL='any string'    */
          /* MSG_LEVEL= not even coded as an option */
                  ASCII_DOTS='YES'
                  CURRENCY_SYMBOL='$');
proc print data=asia_only;
var sales;
run;
 ods excel close;
```

SAS Log 5-1 – Log Output from MSG_LEVEL Suboption Testing

```
34    ods excel file="&path.\MSG_LEVEL_No_in_Workbook.xlsx"
35          options(
36               /* MSG_LEVEL='NO'                */
ERROR: Option "ascii_dots" not recognized.
ERROR: Option "currency_symbol" not recognized.
NOTE: The EXCELXP option ascii_dots is not supported in this Excel
tagset.
NOTE: The EXCELXP option currency_symbol is not supported in this
Excel tagset.

37               /* MSG_LEVEL='YES'               */
38               /* MSG_LEVEL='any string'    */
39               /* MSG_LEVEL= not even coded as an option */
40                       ASCII_DOTS='YES'
41                       CURRENCY_SYMBOL='$');
42
43    proc print data=asia_only;
44    var sales;
45    run;

NOTE: There were 14 observations read from the data set
WORK.ASIA_ONLY.
NOTE: PROCEDURE PRINT used (Total process time):
      real time              0.18 seconds
      cpu time               0.10 seconds


46
47    ods excel close;
NOTE: Writing EXCEL file: I:\SAS__Book_2016\Chapter 5 Excel
      Output\MSG_LEVEL_No_in_Workbook.xlsx
```

The TAB_COLOR= Option

The TAB_COLOR= option is fun to play with. You have many different ways that you can express a color for the Excel tab. Of course, the colors, tints, shades, and hues might not be displayed the same on all monitors, printers, or even within Excel itself. But the ability to use a color name, hexadecimal value, red-green-blue (RGB) density array, and even add a transparency value to red-green-blue (RGBA) makes the process very flexible. And it offers you the choice of using the Excel pallet or the full 16,777,216 color range. In this example, I used the SHEET_NAME= option to compress the size of the text on the worksheet for viewing.

SAS Code 5-7 – Several Examples of Using the TAB_COLOR= Option

```
ods excel file = "&path\color_tabs.xlsx"
          options(sheet_name='A');
  proc print data=Asia_only;
  run;
ods excel options(tab_color='blue');
  proc print data=Asia_only;
  run;
ods excel options(tab_color='#00ff00');
  proc print data=Asia_only;
  run;
ods excel options(tab_color='rgb(255,00,0)');
  proc print data=Asia_only;
  run;

ods excel options(tab_color='rgba(0,100,20,0.5)');
  proc print data=Asia_only;
  run;
ods excel options(tab_color='rgb(0,75%,50%)');
  proc print data=Asia_only;
  run;

ods excel options(tab_color='rgba(80%,0%,0%,0.75)');
  proc print data=Asia_only;
  run;

ods excel close;
```

Figure 5-8 – Several TAB_COLOR= Option Outputs

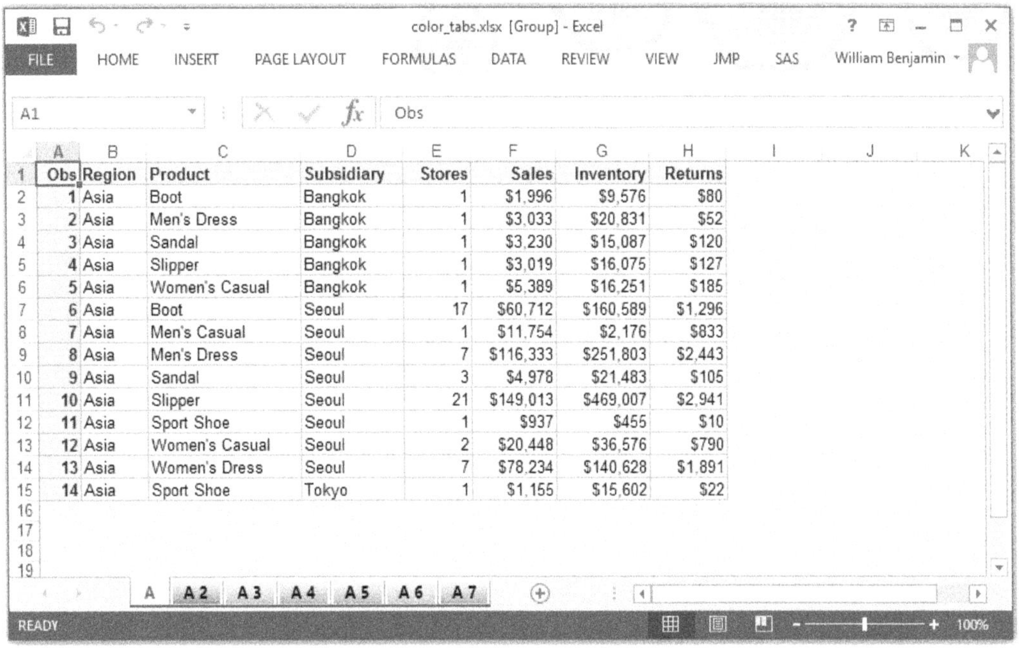

The ZOOM= Option

When the data in a worksheet will not fit on your screen, you can reduce the size of the text by using the slider in the lower right of the screen. This enables you to see more columns. The ZOOM= option can provide the same feature, but programmatically. Below I reduced the size of the text of the worksheet but showed the workbook screen print in the same scale as the other images in the book. The data in the image is not meant to be readable, this is just an example showing it can be done. Note the slider in the lower right says 25%. The SCALE= option takes care of the printed output while the ZOOM= option is for the nonprinted output.

SAS Code 5-8 – Using the ZOOM= Option to Reduce the Image

```
ods excel file="&path.\Zoomed_Workbook.xlsx"
          options(zoom='25');
proc print data=asia_only;
run;
ods excel close;
```

Figure 5-9 – Output of the ZOOM= Option

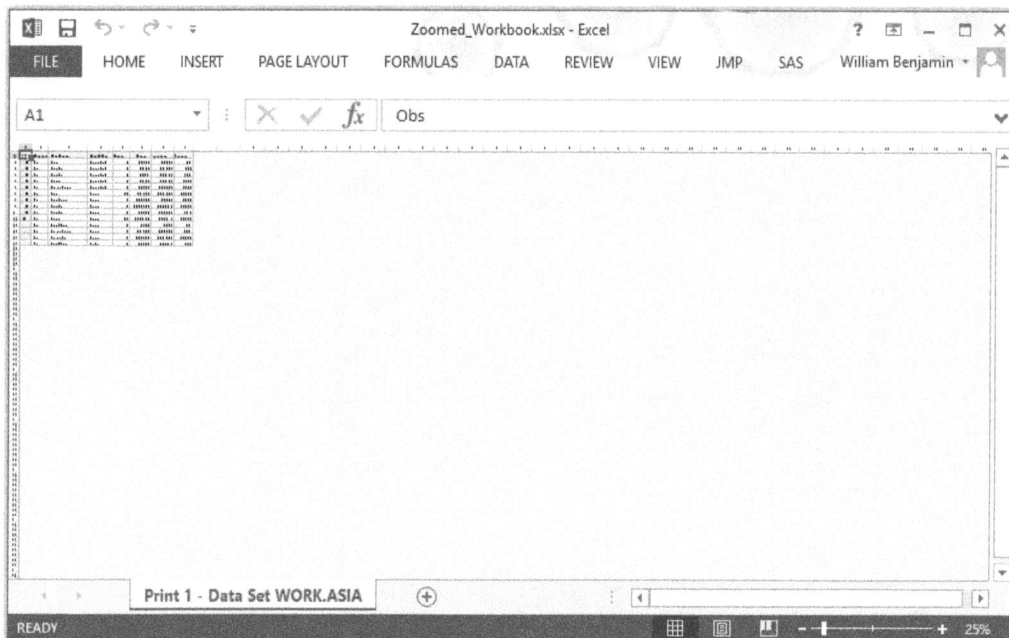

Conclusion

As you can see the options shown in this chapter allow the programmer to preset many features that are time consuming to reproduce within the Excel file manually. I have shown how several different output files and file types can be produced in one pass over the SAS code, cutting down on the program execution time and manual errors. The manual generation of a Table of Contents can take hours to locate and to create the hyperlinks. The feature to preset tab colors and zoom level are also helpful for data identification and positioning on the screen or printed page.

Chapter 6: Arguments that Affect Output Features

Introduction

In this chapter I will be discussing and showing examples that impact the appearance of the final output Excel worksheet. This includes SAS Style options, Cascading Style Sheets, and options that affect the location, density, and size of text and images in the output Excel worksheets. These options control worksheet-level features of the output Excel file.

Using these options, you can modify the way the basic colors of the spreadsheet appear, and the size of the cells in the output worksheet. Text can be added to the worksheet, anchors for hyperlinks can be added, graphical image density can be modified, default SAS styles can change the worksheet, and external cascading style sheets can be applied. All of these features can be added manually after the workbook is created, but these options enable you to make the desired output when the workbook is created. This allows the workbook to be displayed the way that you want it to appear and have the creation and last modified date be the same. Again, the SAS data set "ASIA_ONLY" is defined in Chapter 2.

Arguments That Affect Worksheet-Level Output Features

In this chapter we will discuss the following SAS ODS Excel destination topics. I consider these topics to affect the overall appearance of the output worksheet.

Table 6-1 – ODS Excel Destination Actions

Action Parameter / Alias	Options	Description
DOM	File="quoted external filename" or Blanks meaning no defined file name.	This option writes the output from the document object model to either an external file or the SAS Log if no file name is supplied. *
ANCHOR	'anchor-name'	Specifies the anchor tag name, 'IDX' is the default first anchor.
BOX_SIZING	CONTENT_BOX or BORDER_BOX	This option describes how to measure the width of the cells (this can also be adjusted by using the SAS Registry**).
CSSSTYLE	'file-specification'<(media-type1<...media-type-10>)>	Fully qualified name of a Cascading Style
IMAGE_DPI/DPI	'number'	Dots per inch to use for graphical output
STYLE	style-override(s)	Use a predefined style element (a collection of style changes) or a single (or group of) style name-value pair of changes.
TEXT	text-string	Inserts a text string into the output document.

* For more information about the ODS document object model (DOM), see the following URL: http://support.sas.com/documentation/cdl/en/odsadvug/69833/HTML/default/viewer.htm#n0pg2pc6qv5vzhn160f8j7ap4uv 1.htm or the SAS manual "SAS Institute Inc. 2016. *SAS® 9.4 Output Delivery System: Advanced Topics, Third Edition.* Cary, NC: SAS Institute Inc."

** Modification of the SAS Registry is not discussed in this book.

Using the ODS Document Object Model (DOM) with the ODS EXCEL Statement

The SAS documentation describes the ODS Document Object Model or DOM as one of the options available to the ODS EXCEL statement. Actually, the DOM option can be applied to most ODS statements except the ODS LISTING and ODS OUTPUT statements. The *SAS 9.4 Output Delivery System: Advanced Topics, Third Edition* manual noted above says that the DOM feature of ODS allows you to "determine what elements and attributes exist so that you can construct your CSS selectors to access those areas." Again, space and time do not permit me to give a full explanation of this ODS feature. When used with the ODS EXCEL statement two formats are available.

1. ODS EXCEL DOM;
2. ODS EXCEL DOM <="external-file"> ;

Statement number 1 writes the output to the SAS Log, while statement number 2 writes to an external file. But what does the output look like? SAS Code 6-1 shows the syntax of the first format, and SAS Log 6-1 shows a partial listing of the output. Format 2 just writes the same information to an external file and is not shown here. I also used PROC PRINT to show output information that the SAS Log 6-1 displays. The ODS TRACE statement is used here to describe the ODS outputs generated during the code execution. The ODS TRACE statement is briefly described in Chapter 3.

SAS Code 6-1 – Using the DOM Option to Identify CSS Elements

```
ods trace dom;
ods excel dom;

ods excel options(sheet_interval='none');

proc print data=sashelp.shoes(where=
      (Product     = 'Boot'      and
       Subsidiary = 'New York' and
       region      = 'United States'));
run;

ods excel close;
ods trace off;
ods html;
```

SAS Log 6-1 is only a partial listing of the output. It is shown here to help identify some of the elements used later by the ODS EXCEL ANCHOR= option.

SAS Log 6-1 – Partial Listing of the DOM Trace Output

```
296  ods trace dom;
297  ods excel dom;

. . . more dom trace output . . .

298
299  ods excel options(sheet_interval='none');
300
301  proc print data=sashelp.shoes(where=
302        (Product     = 'Boot'      and
303         Subsidiary = 'New York' and
304         region      = 'United States'));
305  run;

        <div>
        </div>
        <section id="idx" class="oo" data-name="procprinttable"
label="data set
sashelp.shoes" proc="print" output="print" contents-label="data set
sashelp.shoes">

. . . more dom trace output . . .

                          <th class="header" type="char"
unformatted-type="char" index="2"
name="region" data-name="region" label="region">Region
```

```
                                        </th>
                                        <th class="header" type="char"
unformatted-type="char" index="3"
name="product" data-name="product" label="product">Product
                                        </th>
                                        <th class="header" type="char"
unformatted-type="char" index="4"
name="subsidiary" data-name="subsidiary"
label="subsidiary">Subsidiary
                                        </th>

. . . more dom trace output . . .

type="char" index="2"
name="region" data-name="region" label="region">United States
                                        </td>
                                        <td class="data" type="char" unformatted-
type="char" index="3"
name="product" data-name="product" label="product">Boot
                                        </td>
                                        <td class="data" type="char" unformatted-
type="char" index="4"
name="subsidiary" data-name="subsidiary" label="subsidiary">New York
                                        </td>
                                        <td class="data" type="num" unformatted-

. . . more dom trace output . . .

NOTE: There were 1 observations read from the data set
SASHELP.SHOES.
        WHERE (Product='Boot') and (Subsidiary='New York') and
(region='United States');
NOTE: PROCEDURE PRINT used (Total process time):
        real time           0.32 seconds
        cpu time            0.28 seconds

306
307  ods excel close;

. . . more dom trace output . . .

NOTE: Writing EXCEL file: .\sasexcl.xlsx
308  ods trace off;
309  ods html;
NOTE: Writing HTML Body file: sashtml10.htm
```

The bold highlighted text phrases section id="idx", class="header", and class="data" will be expanded upon when explaining the ODS Anchor option. These are CSS elements that the ODS EXCEL DOM option and the ODS TRACE DOM statement showed. Of course, many others exist and in fact at least four other class statements exist in each of the "header" and "data" parts of the listing. The ODS TRACE listing had over 7 pages of listing, far too much data to display. Table 6-2 includes some of the element types available, as well as a brief definition. For more information, see SAS Institute Inc. 2014. *SAS® 9.4 Output Delivery System: Advanced Topics, Third Edition*. Cary, NC: SAS Institute Inc.

Table 6-2 – CSS Sector Classes

A Limited Selection of CSS Selector Classes		
Selector Class	**Description**	**Example**

Class Selectors	Class selectors are style selectors that select elements based on the value of the class= attribute in the markup of an ODS report. Class selectors must have a period (.) preceding the class name. For example, in the following rule set, the class style selector is .SYSTEMTITLE.	.systemtitle
		{ font-family: arial, helvetica, sans-serif;
		  color: red;
		  border: 1px solid black;
		}

Element Selectors	Element selectors are style selectors that select DOM elements based on the element name. For example, the following rule set selects elements with the name P:	p {color:green}
Universal Selector	The universal selector is a style selector that is a wildcard. It can match any element name. The syntax for the universal selector is an asterisk (*).	*
Pseudo-Class Selectors	Pseudo-class selectors are style selectors that select elements based on the relationships between DOM elements. Pseudo-classes are represented by the pseudo-class name prefixed with a colon (:). The following are examples of some ways that you can use pseudo-class selectors: • select the first and last child of a parent element • select a specific child based on its positional index in the parent element • select an element by position of a particular element name	:root selects the top-level element in the DOM. :first-child selects the first element within the parent. :first-of-type selects the first element of that type (that is, same element name) in the parent. :marker selects a list item bullet in printer output. :nth-child(an+b) selects an element based on the equation an +b. This equation selects every *a*th element starting with element at position *b*. The equation can be replaced with the keywords even or odd for the simple case of alternating the selection. :nth-of-type(an+b) selects the :nth-child, except that only the same elements of the same type are used in the calculation. :empty selects one or more empty

A Limited Selection of CSS Selector Classes		
Selector Class	**Description**	**Example**
		elements. This only applies to elements that have been specified as empty by the procedure.
		:before inserts content before the element.*
		:after inserts content after the element* :not(…) selects an element if the selector within the argument is not true.
ID Selectors	ID selectors are style selectors that select elements based on the id= attribute of a DOM element. The ID must be unique within a DOM and only one can be specified in the id= attribute. ID selectors are indicated by a "#" prefix. The following is a CSS rule set with an ID selector:	#idx1 { font-style: italic }
Attribute Selectors	Attribute selectors select DOM elements with the specified attribute. ID selectors and class selectors are special case attribute selectors. Attribute selectors use the following syntax to select attributes: [*attribute operator "value"*] Operator specifies the operator. Operator allows partial matches.	= matches the entire attribute value. ^= matches the beginning of an attribute value. =$ matches the end of an attribute value. *= matches any substring in an attribute value. ~= matches any space-separated word in an attribute value. This operator can be used to emulate the class selector. |= matches an attribute value and an optional value followed by a hyphen. This operator is used to match language codes such as enUS, en-GB, and so on.
Combinators	Combinators are characters that select an element based partially on its context within another element. This is done by combining selectors using one of the following characters.	" " (space) indicates that the selector to the left must match an element anywhere in the parentage of the currently selected element. > selects elements that are a direct descendent of the specified element. ~ selects elements that have another sibling anywhere within the parent.

A Limited Selection of CSS Selector Classes		
Selector Class	Description	Example
		+ selects elements that have a specified element immediately preceding them

* This is a special case that is used the same way as the PRETEXT, PREIMAGE, POSTTEXT, and POSTIMAGE style attributes.

Figure 6-1 – Excel Output Using the DOM Option to Identify CSS Elements

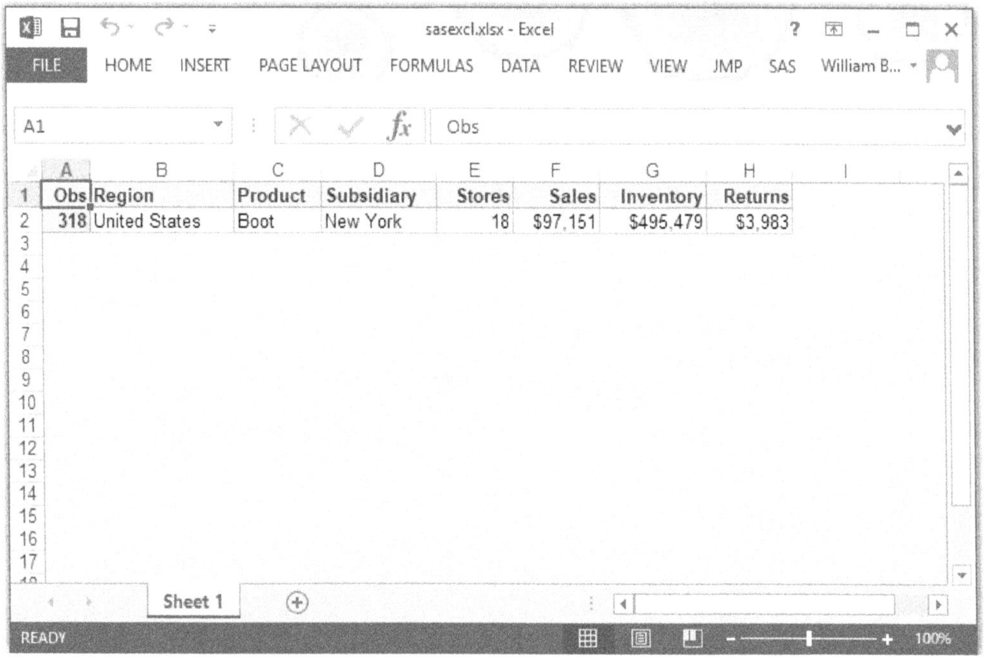

Figure 6-1 is an image of the output of example SAS Code 6-1. It looks pretty simple. There are no surprises here. The code in SAS Code 6-1 did not apply any style changes to the workbook. It only restricted the number of rows processed so that the DOM trace output could be small enough to allow the identification of a few CSS elements for use later.

The ODS ANCHOR= Option

An Anchor tag is an important part of the ODS output. You can specify an "ANCHOR" or allow ODS to generate them using default names. The syntax is to provide the keyword ANCHOR and an anchor-name, as follows: ANCHOR='anchor-name'. The default anchor name is id="idx". Once you specify the first anchor, that name is used and a number is added to the end of the name for the next anchor. The code in SAS Code 6-2 created a user generated Cascading Style Sheet with two references called "#Expense .header" and '#Expense .data". Each of these references modify some portion of the Excel output worksheet colors. When you use an ANCHOR, you are changing the ID= attribute of the current style sheet output in use, this is effective in both SAS style sheets and Cascading style sheets (#ID=). To identify where these anchors appear in the output style, use the DOM option. An example of how to use the ODS EXCEL DOM option is described above, see SAS Code 6-1.

In this section I will provide a simple example. There are many documents, papers, and books written on this subject in which you can find more examples. The official website of the World Wide Web Consortium that maintains the CSS standards is https://www.w3.org/Style/CSS/.

Another good source of information about this type of data manipulation is a SAS Global Forum paper found at the following URL:
http://support.sas.com/resources/papers/proceedings15/SAS1880-2015.pdf

SAS Code 6-2 – Using a Style Sheet to Modify Excel Output.

```
* Build a Cascading Style Sheet file;
filename my_css "css_text.css";
data _null_;
   file my_css noprint notitles linesize=256;
   put
      '#Expense .header {background-color:orange}' /
      '#Expense .data   {background-color:pink  }' ;
run;
Ods html close;

* Use a Cascading Style Sheet file;
ods excel cssstyle = "css_text.css"
         options(sheet_interval='none');
* Use an ANCHOR;
ods excel anchor="expense";
proc print data=sashelp.shoes(where=
      (Product     = 'Boot'       and
       Subsidiary  = 'New York' and
       region      = 'United States'));
run;
ods excel close;
ods html;
```

The CSS file created in SAS Code 6-2 modified the background colors to Orange and Pink. The ODS EXCEL CSSSTYLE= option is described later in this chapter.

Figure 6-2 – Excel Output Using the ANCHOR= option and a CSS File to Change Excel Output

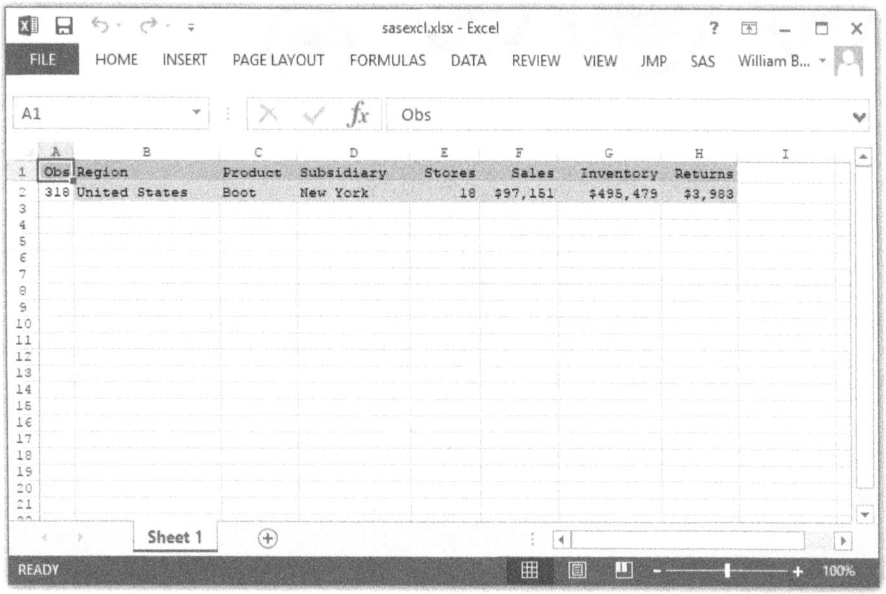

Notice that Column 'A' Row '2' is not a "DATA" element, it is a ROW HEADER.The CSS file does not contain background color instructions for ROW HEADER; therefore, it is given the default background color of white.

The BOX_SIZING= Option

The output of this option is hard to see, I will show the code here for one of the options, but not the output because the difference from the default output is not really noticeable at the resolution available in this document. This option interacts with the borders of the cells of the output Excel worksheet. I found that the differences are as small as one or two pixels and the width of the boarders are only noticeable at very large magnifications. It is also closely tied to Cascading Style Sheet processing. The SAS Documentation points to the following WC3/CSS3 website for a more detailed description. http://www.w3.org/TR/2002/WD-css3-box-20021024/#box-sizing. However, the syntax shown on the web page does not seem to apply to the SAS syntax. The web site basically says that the difference is that the CONTENT_BOX= option derives the width from the content of the cell, while the BORDER_BOX= option will derive the width of the cell from calculations that include the width of the border and the padding. Both of these definitions are expressed (visible) as changes in the number of pixels used to print the border of an Excel data cell.

SAS Code 6-3 – The BOX_SIZING= Option

```
ods excel file = "&path\box_sizing_border_box.xlsx"
          box_sizing=border_box;
   proc print data=Asia_only;
   run;

ods excel close;
```

Using Cascading Style Sheets with the CSSSTYLE= Option

SAS and Cascading Style Sheet usage is so pervasive, even I wrote a paper on the topic [1]. I showed how to create and include a simple CSS file into Excel using the ODS Excel destination. To save you from having to look up that paper I will include some of the information here.

[1] Benjamin, William E, Jr. Owl Computer Consultancy, LLC, Phoenix AZ, 2016. "Working with the SAS® ODS EXCEL Destination to Send Graphs, and Use Cascading Style Sheets When Writing to EXCEL Workbooks" Published on the WEB site http://www.mwsug.org/proceedings/2016/HW/MWSUG-2016-HW04.pdf.

Simple Color Changes Using a CSS File

This section will provide an example of making simple color changes to the output Excel worksheet "body", "header", "rowheader", and "data" portions of the worksheet using a CSS file generated in a few lines of SAS code.

SAS Code 6-4 – Creating a Cascading Style Sheet File and a Simple SAS Work File

```
 * Define a CSS output file;
filename css "&path.\my_css_1.css";

* Write a css file;
data _null_;
file css noprint linesize=132;
put
".body  {background-color: lightblue;  } " /
".header{background-color: gold;       } " /
".rowheader {background-color: purple;    " /
"           color: white;              } " /
".data  {background-color: lightgreen; } " ;
 run;

 * Create a simple file to write to Excel;
 options linesize=255;
 data test_css;
     Column_a =  'My test item 1'; output;
     Column_a =  'My test item 2'; output;
     Column_a =  'My test item 3'; output;
     Column_a =  'My test item 4'; output;
     Column_a =  'My test item 5'; output;
     Column_a =  'My test item 6'; output;
     Column_a =  'My test item 7'; output;
 run;
```

The file "my_css_1.css" is a cascading style sheet that changes the color of five different sections of the output Excel workbook. The SAS data set "test_css" is simply seven character strings that mean nothing. Both of these files are used in SAS Code 6-5 and SAS Code 6-6.

SAS Code 6-5 – Create an Excel Workbook without Using a Cascading Style Sheet

```
* Write a file to EXCEL without CSS changes;
ods excel file = "&path.\Css_file_1.xlsx";
proc print data=test_css;
run;
ods excel close;
```

This code results in a simple Excel workbook, as shown in Figure 6-3.

Figure 6-3 – An Excel Workbook without Using a Cascading Style Sheet

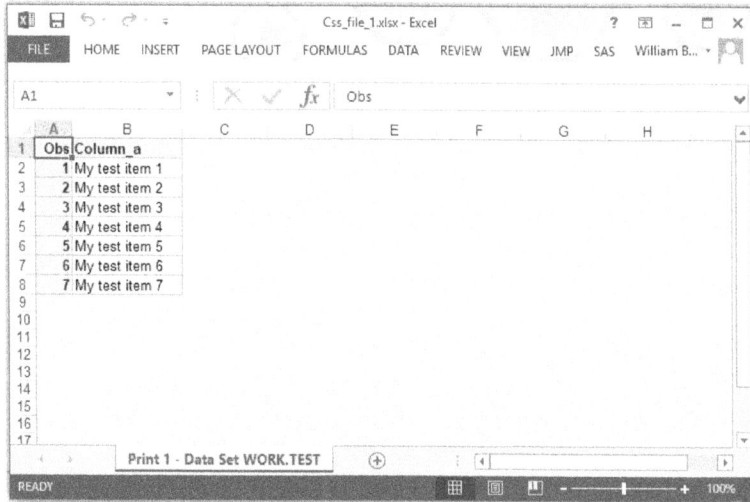

No special features were applied to the workbook created in Figure 6-3.

SAS Code 6-6 – Create an Excel Workbook Using a Cascading Style Sheet

```
* Write a file to EXCEL with CSS changes;
ods excel  file  = "&path.\Css_file_2.xlsx"
        cssstyle = "&path.\my_css_1.css";
proc print data=test_css;
run;
ods excel close;
```

Figure 6-4 An Excel Workbook Using a Cascading Style Sheet

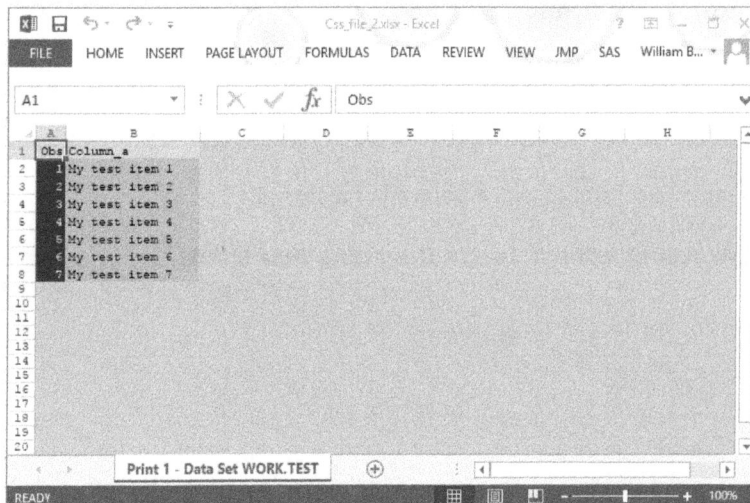

Colors can make a real difference, but choose your own and they perhaps will look better.

Sending Graphs to Excel and Modifying the Output Using a CSS File

One of the things that you will find out in life is that by learning a lot you will figure out that there is still a lot more to learn. This book was never intended to teach you everything about the ODS Excel options and suboptions. But because of the size and prevalence of CSS, I would be remiss if I did not show an example of adding a graph and an image to an Excel spreadsheet and modifying it using the ODS CSSSTYLE= option.

First, let's create a CSS file to make some magical color changes to a graphical output image.

SAS Code 6-7 – CSS File to Change Colors on a Graph

```
filename my_css "my_css_2.css";

data _null_;
   file my_css noprint notitles linesize=132;
   put
      ".graphbackground       {background-color: orange;     } " /
      ".graphtitle1text       {color: blue; Font: 20pt arial; " /
      "                        font-weight: bold;            } " /
      ".graphfootnotetext     {color: blue; Font: 12pt arial; " /
      "                        font-weight: bold;            } " /
      ".graphvaluetext        {color: black;                } " /
      ".graphdata1            {color: red;                  } " /
      ".graphdata2            {color: white;                } " /
      ".graphdata3            {color: blue;                 } " /
      ".graphdata4            {color: red;                  } " /
      ".graphdata5            {color: white;                } " /
      ".graphdata6            {color: blue;                 } " /
      ".graphdata7            {color: red;                  } " /
      ".graphdata8            {color: white;                } " /
      ".graphdata9            {color: blue;                 } " ;
   run;
```

Now we need to have a way to reproduce a graph so that it can be displayed in two Excel workbooks.

SAS Code 6-8 – A Macro to Generate a Graph

```
%macro Graph_it;
    PROC SQL;
    CREATE VIEW WORK.Sorted_1 AS
        SELECT T.Region, T.Sales
            FROM SASHELP.SHOES(WHERE=(Region ne "Asia")) as T;
    QUIT;

    Legend1
        FRAME
        POSITION = (BOTTOM CENTER OUTSIDE);

    TITLE1 "Pie Chart of SASHELP.Shoes Return data by Region";
    TITLE2 "Execpt Asia";
    FOOTNOTE1 "Produced by William E Benjamin Jr";

    PROC GCHART DATA =WORK.Sorted_1;
        PIE3D  Region / SUMVAR=Sales
        TYPE=SUM            LEGEND=LEGEND1
        SLICE=OUTSIDE       PERCENT=OUTSIDE
        VALUE=OUTSIDE       OTHER=4
        OTHERLABEL="Other"  COUTLINE=BLACK
        NOHEADING;
    RUN;
    QUIT;

    TITLE;
    FOOTNOTE;
    RUN;
%mend Graph_it;
```

SAS Code 6-9 – Outputting a Graph to Excel without CSS Modifications

```
/* without CSS style modification */
ods excel file = "&path.\Test_css_file_1.xlsx";
%graph_it;
ods excel close;
```

This code executes the graph using a macro to produce a graph, no CSS modifications are applied and the output is based on default style template that is used.

Figure 6-5 – An Excel Graph without CSS Modifications

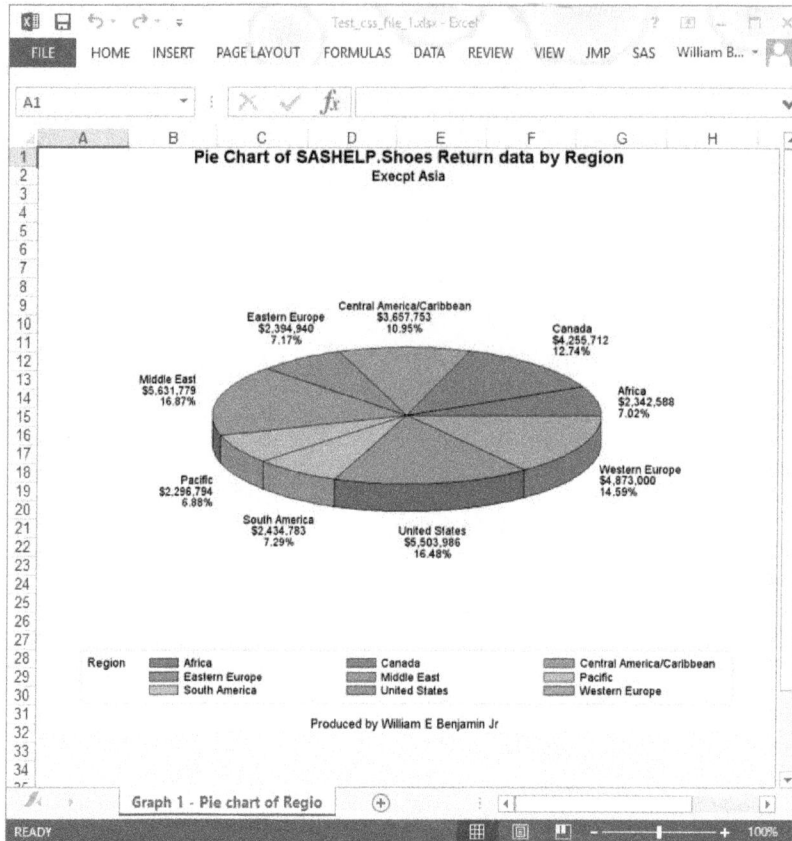

Every part of the Graph in Figure 6-5 has its own color.

SAS Code 6-10 – Outputting a Graph to Excel with CSS Modifications

```
/* with CSS style modification */
ods excel file = "&path.\Test_css_file_2.xlsx"
    cssstyle = "&path.\my_css_2.css";
%graph_it;
ods excel close;
```

The CSS file had a big affect on the output graph. It is much more colorful, but much less readable. I never said that all CSS file commands made good changes to an output process. But hopefully these code snippets will give you something that you can use to figure out what elements of the CSS file changed your graphical output.

Figure 6-6 – An Excel Graph with CSS Modifications

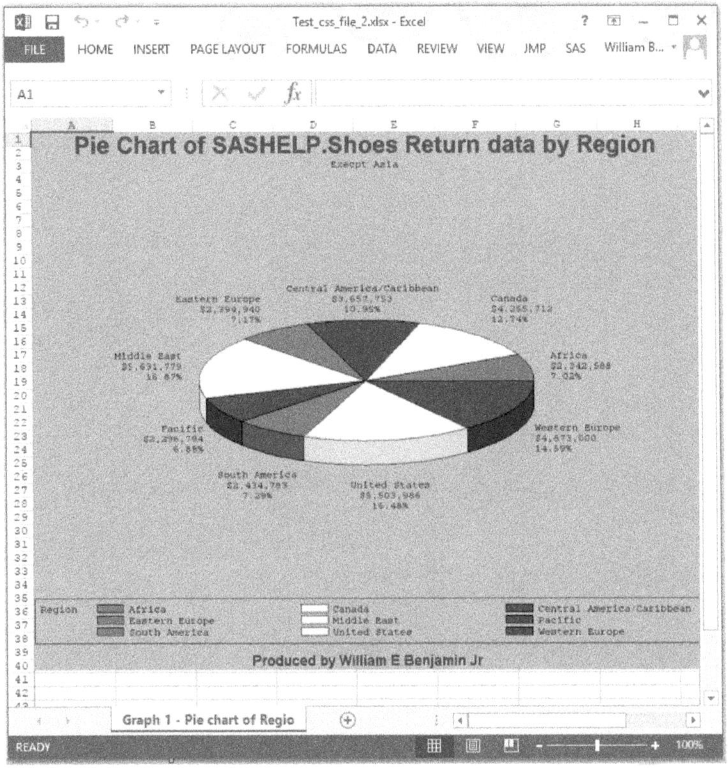

File "my_css_2.css" has been applied to the SAS output when creating the Excel workbook.

Adding a Background Image Using CSS

CSS Files and URLs are used in a wide variety of applications, not just Excel workbooks. It is neat to know that now you can easily use not only graphical images generated by SAS, but also photographic images when you want to spruce up your spreadsheets. In the example in SAS Code 6-11 I use the same graph as in Figure 6-6 and add an image into the background. I have deliberately chosen an image bigger than it needs to be. I shrank the graph image so that nearly the whole picture of the horse's face would be visible. I did this to show that the images repeat themselves. A faint batch of brown can be seen on the right side of the image. This is the image starting to repeat.

SAS Code 6-11 – Building a CSS File to Include a URL Reference

```
filename my_css "&path.\my_css_3.css";
* See SAS Code 6-0 for the definition of &path;
* the default is "I:\SAS__Book_2016\Chapter 6 Excel Output";
* when downloaded the file Wild_Horse_in_AZ.tif is in with the;
* Chapter 6 SAS code,it needs to be moved to the &path directory;

data _null_;
   file my_css noprint notitles linesize=132;
   put
     ".body                      {background-image :            " /
```

```
"                             url(&path\Wild_horse_in_AZ.tif);}  " /
".graphbackground            {background-color: orange;        }  " /
".graphtitle1text            {color: blue;  Font: 20pt arial;     " /
"                             font-weight: bold;               }  " /
".graphfootnotetext          {color: blue;  Font: 12pt arial;     " /
"                             font-weight: bold;               }  " /
".graphvaluetext             {color: black;                    }  " /
".graphdata1                 {color: red;                      }  " /
".graphdata2                 {color: white;                    }  " /
".graphdata3                 {color: blue;                     }  " /
".graphdata4                 {color: red;                      }  " /
".graphdata5                 {color: white;                    }  " /
".graphdata6                 {color: blue;                     }  " /
".graphdata7                 {color: red;                      }  " /
".graphdata8                 {color: white;                    }  " /
".graphdata9                 {color: blue;                     }  " ;
run;
```

The URL to add an image to the Excel workbook is added to this code, with comments about where to find and place the image.

SAS Code 6-12 – Using a CSS File to Include a URL Reference

```
/* with CSS style modification */
ods excel file = "&path.\Test_css_file_3.xlsx"
    cssstyle = "&path.\my_css_3.css"
    options(zoom='15');
%graph_it;
ods excel close;
```

Make sure you look at SAS Code 6-11 to verify the location pointed to by the macro variable &PATH. Then adjust the filenames and image names so that you can read in the image that you want. With a smaller image the repeating pattern will be more noticeable. In Figure 6-7 I made the graph very small, 15 percent of the normal size to show off the image.

Figure 6-7 – An Excel Graph with CSS and a URL Added

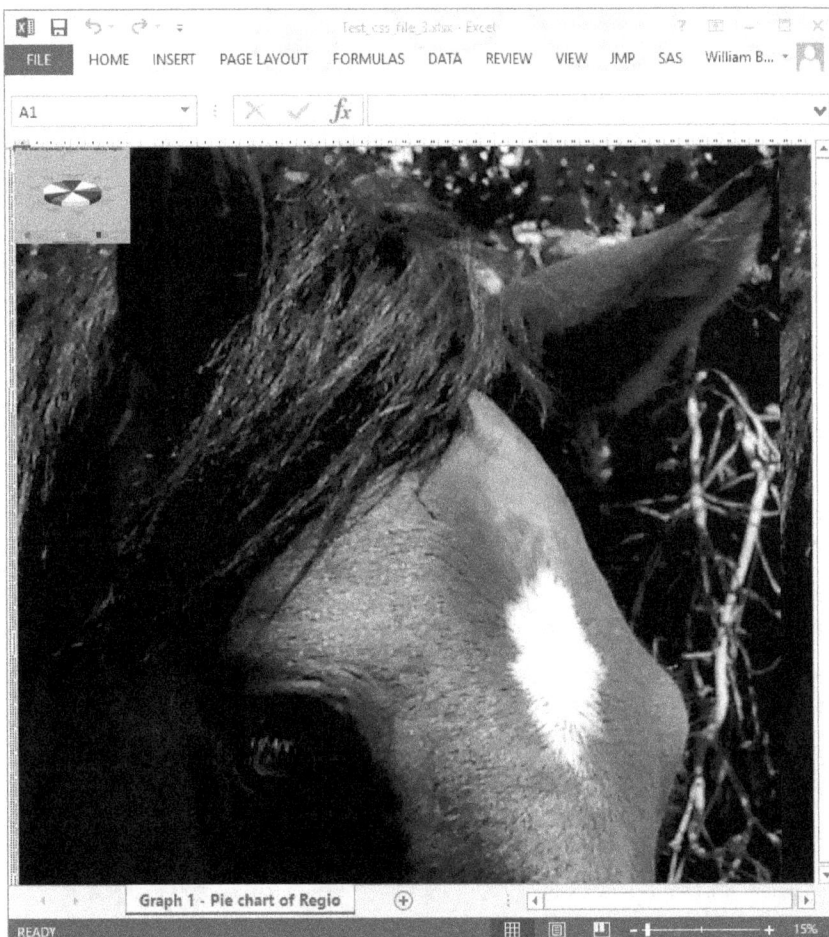

The IMAGE_DPI= Option

This ODS EXCEL option enables you to adjust the pixel density of the output image. An image with more pixels to the inch can be expanded and not lose clarity as fast as a less dense image. To show the effect of this option I have produced two images with different IMAGE_DPI values. SAS Code 6-13 creates an image with the default IMAGE_DPI density of 150 pixels per inch. While SAS Code 6-14 code uses an IMAGE_DPI of 700. Figures 6-8 and 6-9 are both expanded to 400 times normal when viewed with Excel.

Be careful, when the IMAGE_DPI is too large you can run out of memory and might need to resort to increasing the SAS MEMSIZE= system option to a value over 2GB if you run out of memory.

SAS Code 6-13 – Create an Excel Graph with Default IMAGE_DPI Setting

```
ods excel file = "&path.\Test_file_5.xlsx";
%graph_it;
ods excel close;
```

Figure 6-8 – An Excel Graph with Default IMAGE_DPI Setting

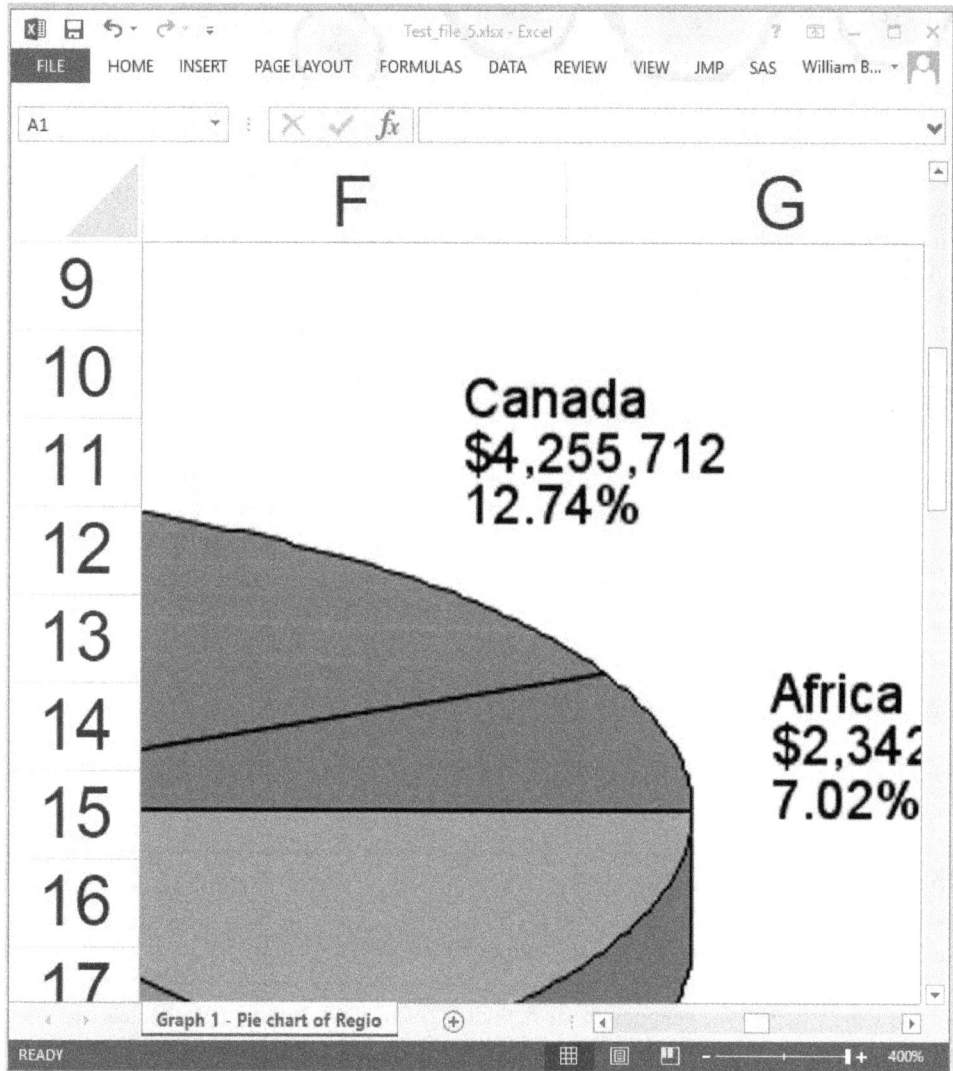

Notice the rough edges on the Pie Chart, and the downward stroke of the '7' and '2'.

SAS Code 6-14 – Create an Excel Graph with IMAGE_DPI=700 Setting

```
ods excel file = "&path.\Test_file_6.xlsx"
          image_dpi='700';
%graph_it;
ods excel close;
```

Figure 6-9 – An Excel Graph with an IMAGE_DPI = 700 Setting

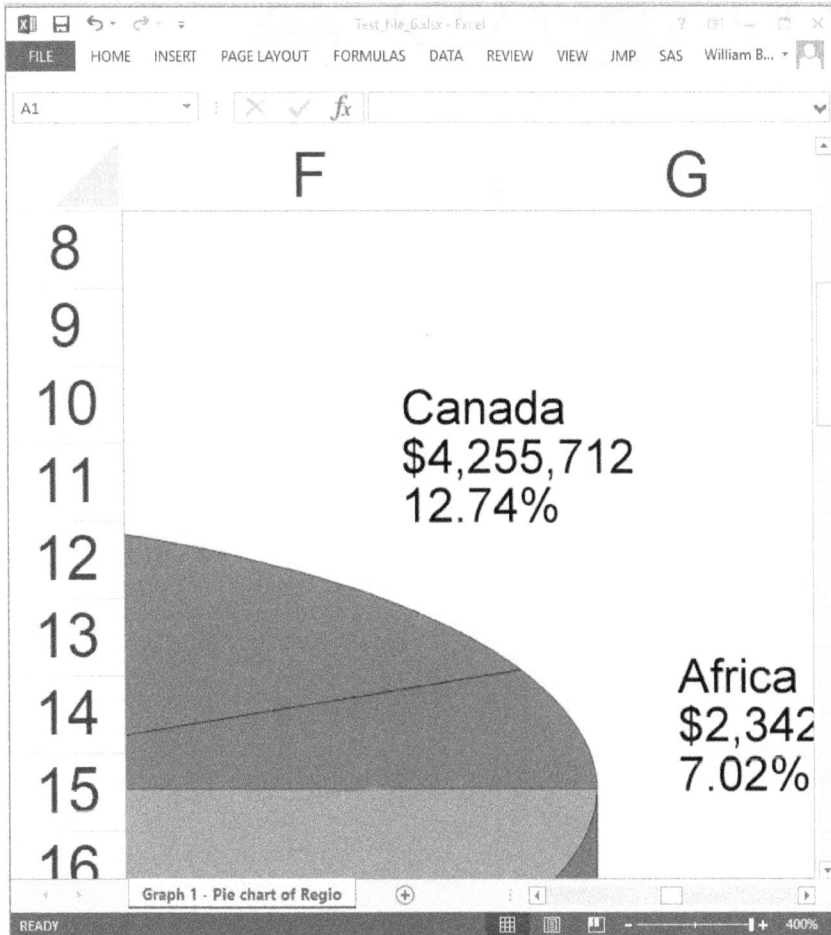

See the clear lines on the edge of the pie chart, and the '7' and '2' no longer look fuzzy.

The STYLE= Option

Hidden beneath each ODS output we use is a default style. When the ODS EXCEL statement is used to write an Excel workbook there is always a style used. The default STYLE is EXCEL. The ODS EXCEL STYLE= option enables you to modify that default.

The ODS EXCEL STYLE= Option

There is a way to determine what styles are available in your current version. SAS Code 6-15 generates a list of the available styles. They are displayed by PROC TEMPLATE. They reside in the SASHELP.TMPLMST item store. The Table of Supported SAS Styles below contains the names of the styles supported in SAS version 9.4 1M3. The SAS code prints a list and I copied the list into the table shown here.

SAS Code 6-15 – Generate a List of SAS Supported Styles

```
ods _all_ close;
ods listing;
   proc template;
      list styles;
   run;
quit;
```

Table of Supported SAS Styles for SAS version 9.4 1M3

List of SAS Styles Supported (SAS 9.4 1M3)			
Analysis	BarrettsBlue	BlockPrint	DTree
Daisy	Default	Dove	EGDefault
Excel	FancyPrinter	Festival	FestivalPrinter
Gantt	GrayscalePrinter	HTMLBlue	Harvest
HighContrast	HighContrastLarge	Journal	Journal1a
Journal2	Journal2a	Journal3	Journal3a
Listing	Meadow	MeadowPrinter	Minimal
MonochromePrinter	Monospace	Moonflower	Netdraw
NoFontDefault	Normal	NormalPrinter	Ocean
Pearl	PearlJ	Plateau	PowerPointDark
PowerPointLight	Printer	Raven	Rtf
Sapphire	SasDocPrinter	SasWeb	Seaside
SeasidePrinter	StatDoc	Statistical	Word
vaDark	vaHighContrast	vaLight	

SAS Code 6-16 – Generate an Excel Workbook with STYLE=SEASIDE

```
ods excel file = "&path.\Test_file_Style_1.xlsx"
              STYLE=SEASIDE;
Proc Print data=Asia_Only;
run;

ods excel close;
```

The default STYLE is EXCEL which produces light blue Column and Row headers. Each of the styles listed in the Table of Supported SAS Styles produces a different layout in the EXCEL workbook. I have not executed code using all of the styles, but I do know that some of the styles only have minor differences from other styles.

Figure 6-10 Excel Workbook Using the SEASIDE STYLE

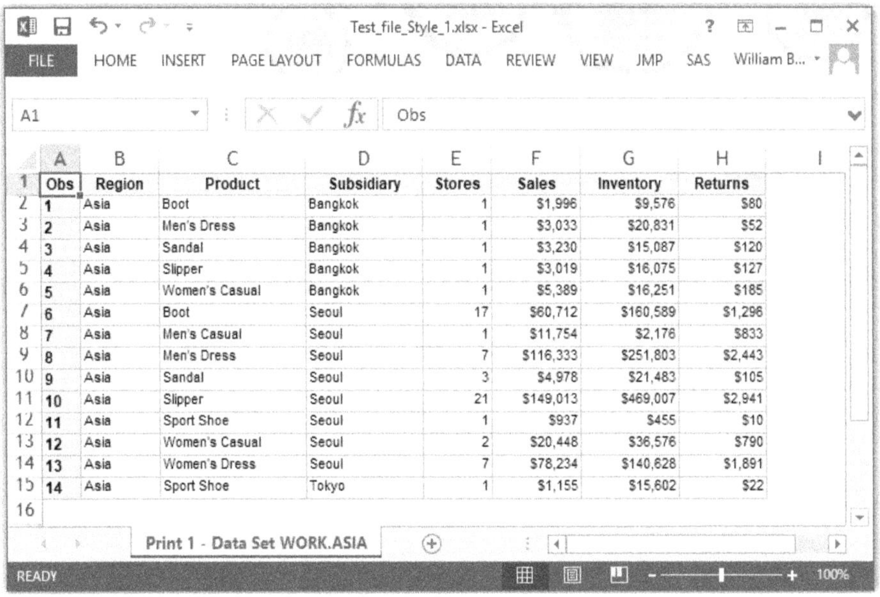

Obs	Region	Product	Subsidiary	Stores	Sales	Inventory	Returns
1	Asia	Boot	Bangkok	1	$1,996	$9,576	$80
2	Asia	Men's Dress	Bangkok	1	$3,033	$20,831	$52
3	Asia	Sandal	Bangkok	1	$3,230	$15,087	$120
4	Asia	Slipper	Bangkok	1	$3,019	$16,075	$127
5	Asia	Women's Casual	Bangkok	1	$5,389	$16,251	$185
6	Asia	Boot	Seoul	17	$60,712	$160,589	$1,296
7	Asia	Men's Casual	Seoul	1	$11,754	$2,176	$833
8	Asia	Men's Dress	Seoul	7	$116,333	$251,803	$2,443
9	Asia	Sandal	Seoul	3	$4,978	$21,483	$105
10	Asia	Slipper	Seoul	21	$149,013	$469,007	$2,941
11	Asia	Sport Shoe	Seoul	1	$937	$455	$10
12	Asia	Women's Casual	Seoul	2	$20,448	$36,576	$790
13	Asia	Women's Dress	Seoul	7	$78,234	$140,628	$1,891
14	Asia	Sport Shoe	Tokyo	1	$1,155	$15,602	$22

Notice the yellow Column and Row headers.

The ODS EXCEL STYLE= Overrides

There are other ways to "Stylize" your output within Excel worksheets, and some of them even have "STYLE=" as part of the name. However, styles are applied within the procedures, not the ODS statement. As a result I will list some of the different types of style overrides, but not show detailed examples. I found these examples on page 311 of SAS Institute Inc. 2016. *SAS® 9.4 Output Delivery System: User's Guide, Fifth Edition.* Cary, NC: SAS Institute Inc.

There are two methods of providing style overrides. First, as a style element, which is a collection of attributes that affect some output of a SAS program. Second, as a style attribute, which is a name-value pair that describes an output behavior or visual result that you want to apply to output data. A style attribute change is the most specific way to directly change how your data looks.

SAS Code 6-17 – Syntax of the Style Overrides

```
/* These code segments are out of context

* The Style-override element name syntax;

style-element-name | [style-attribute-name-1=style-attribute-value-1
<style-attribute-name-2=style-attribute-value-2 ...>]

* The Style-override attribute syntax;

style={tagattr='format:$#,##0_);[Red]\($#,##0\)
formula:RC[-1]-RC[-2]'};

*/
```

These syntax descriptions in SAS Code 6-17 are out of context. These are style overrided but will not execute as coded. In order to get information about the proper way to use these SAS code structures. See SAS Institute Inc. 2016. *SAS® 9.4 Output Delivery System: Procedures Guide, Third Edition.* Cary, NC: SAS Institute Inc. for these and other attribute name-value pairs.

The TEXT= Option

The ODS EXCEL TEXT= option enables you to place a text string into your output. It is that simple.

SAS Code 6-18 – Using the ODS TEXT= Option

```
ods excel file = "&path.\Text_String_1.xlsx"
               options(sheet_interval='NONE');

Proc Print data=Asia_Only (where=(Subsidiary = 'Bangkok'));
run;

ods excel text = ' ';
ods excel text = 'this is two blank lines and a text string in the
middle of my two outputs';
ods excel text = 'the first blank line (row 7) after the first
listing is a space between reports generated by SAS';
ods excel text = ' ';

Proc Print data=Asia_Only (where=(Subsidiary = 'Bangkok'));
run;
ods excel close;
```

There are four ODS TEXT= options shown,

Figure 6-11 – ODS Excel TEXT= Option Output

The ODS TEXT= option lines appear on rows 8 to 11.

Conclusion

The ODS Document Object Model is generated to give information that will help in building the CSS style. This was mentioned in the discussion of the ANCHOR= option. We then moved on to creating Cascading Style Sheet files with commands to change colors, modify graphs, and insert data from a URL. You learned how to make your text and images sharper and clearer by outputting more pixels per inch. Text comments and style overrides rounded up the chapter. This chapter introduces you to many methods and options that you might never have considered using as output to an Excel workbook.

Chapter 7: Options That Affect Worksheet Features

Introduction

This chapter will discuss the ODS Excel options that affect output at the level of the whole workbook. We will discuss titles, footnotes, page alignment, sheet intervals, tab labels, and BYLINE placement. While these ODS Excel options might impact individual worksheets, when left unchanged for the full time that the Excel workbook is being generated, these options have the potential to affect many pages. Therefore, I consider them to affect the whole workbook. Most of these options directly affect the visual display of the worksheets, but the FIT options (FIT_TO_PAGE, PAGES_FITHEIGHT, and PAGES_FITWIDTH) only set options in the Excel workbook. The actual effect of these three options is still under the full control of Excel. This means that Excel might contract or expand the data when printing occurs. Excel also manages the titles and footnotes provided by SAS and this might cause the overlapping of data on printout.

Some of the screen images include the Excel "Page Setup" sheet. When using Microsoft Excel 2013 the Page Setup can be found on the "PAGE LAYOUT" tab under the "Print Titles" option. The "Page Setup" sheet has four tabs; "Page", "Margins", "Header/Footer", and "Sheet".

ODS Excel Destination Suboptions

In this chapter we will discuss the following SAS ODS Excel destination topics.

Table 7-1 – ODS Excel Destination Actions

Suboption Parameter	Values	Description
EMBEDDED_FOOTNOTES	'OFF', 'ON', 'YES', 'NO' Default = 'OFF'	This option specifies whether footnotes should be embedded within the worksheet. Negative options cause footnotes not to be added to the Excel worksheets.
EMBED_FOOTNOTES_ONCE	'OFF', 'ON', 'YES', 'NO' Default = 'OFF'	This option controls whether embedded footnotes appear at the bottom of the worksheet. Positive responses (YES and ON) cause footnotes to appear only once, at the botton of the worksheet. Negative responses (OFF and NO) cause embedded footnotes to appear at the bottom of the output in the worksheet. Option name alias = "EMBED_FOOTERS_ONCE"
EMBEDDED_TITLES	'OFF', 'ON', 'YES', 'NO' Default = 'NO'	This option determines whether the titles appear in the worksheet. Positive responses (YES and ON) will embed the titles while negative responses (OFF and NO) cause titles not to be embedded.
EMBED_TITLES_ONCE	'OFF', 'ON', 'YES', 'NO' Default = 'OFF'	This option controls whether embedded titles appear at the top of the worksheet. Positive responses (YES and ON) cause titles to

Suboption Parameter	Values	Description
		appear only once at the top of the worksheet. Negative responses (OFF and NO) cause embedded titles to appear as they would normally.
FITTOPAGE	'OFF', 'ON', 'YES', 'NO' Default = 'OFF'	This option controls whether Excel should try to print the worksheet onto one page. Positive responses (YES and ON) turn on this option while negative responses (OFF and NO) turn the option off.
GRIDLINES	'OFF', 'ON', 'YES', 'NO' Default = 'OFF'	This option controls whether Excel should try to print gridlines. Positive responses (YES and ON) turn on this option while negative responses (OFF and NO) turn the option off.
PAGES_FITHEIGHT	'number'	Many worksheets contain more data than will print on one sheet of paper. This option specifies how many pages down to fit onto a printed sheet.
PAGES_FITWIDTH	'number'	Many worksheets contain more data than will print on one sheet of papcr. This option specifies how many pages across to fit onto a printed sheet.
SHEET_INTERVAL	'BYGROUP', 'PAGE', 'PROC', 'NONE', 'TABLE', Default = 'TABLE'	This option determines when a new worksheet is created. <ul><li>'BYGROUP' = a new worksheet is created for each new BYGROUP. (Alias BYGROUPS)</li><li>'PAGE' = a new worksheet is created for each new page of procedure output.</li><li>'PROC' = a new worksheet is created for each new procedure and it contains all data for that procedure.</li><li>'NONE' = all of the output is on one page.</li><li>'TABLE' = a new worksheet is created for each table. (Alias OUTPUT)</li></ul>

Suboption Parameter	Values	Description
SHEET_LABEL	'text-string', 'NONE', Default = 'NONE'	If supplied, the 'text-string' is used as the first part of the worksheet name, which replaces the predefined label string. This is a label that prepends to the sheet name. This option can also be used in conjunction with other options like 'SHEET_INTERVAL.' When 'NONE' is specified then the default sheet label is used.
SHEET_NAME	'text-string'	When provided, the text-string is used as the name of the next worksheet. The text-string can be up to 28 characters long. When required, a worksheet counter is added to the worksheet name to ensure the names are unique.
SUPPRESS_BYLINES	'OFF', 'ON', 'YES', 'NO' Default = 'OFF'	Positive responses suppress the output of BY lines in the worksheet, while negative responses allow the BY lines to appear.

The EMBEDDED_FOOTNOTES= Suboption

The EMBEDDED_FOOTNOTES= suboption shown here sends the SAS footnotes to the output Excel worksheet. This is often used with the EMBEDDED_TITLES= suboption to place the SAS titles and footnotes onto the Excel worksheet pages. When multiple pages are produced the footnotes appear on each page. The footnotes can be changed and the new footnote will appear on the next output worksheet. When multiple outputs are sent to one worksheet, the footnotes can be changed between outputs on the same worksheet. The default value or the EMBEDDED_FOOTNOTES='OFF' (or 'NO') results in no footnote being output.

Embedded Footnotes Turned Off

When EMBEDDED_FOOTNOTES are turned off (the default), then no footnotes sent from SAS are displayed on the Excel worksheet pages.

SAS Code 7-1 – Using the EMBEDDED_FOOTNOTES='OFF' Option

```
** Code to generate the Excel file in Figure 7-1 ***;
Footnote 'foot note one';
ods excel file = "&path\Footnote_off.xlsx"
          options(EMBEDDED_FOOTNOTES='off');
   proc print data=Asia_only;
   run;
ods excel close;
```

In SAS Code 7-1 the SAS code has a footnote, but the output worksheet Figure 7-1 shows no SAS footnotes.

Figure 7-1 – Excel Output with the EMBEDDED_FOOTNOTES= Option Off

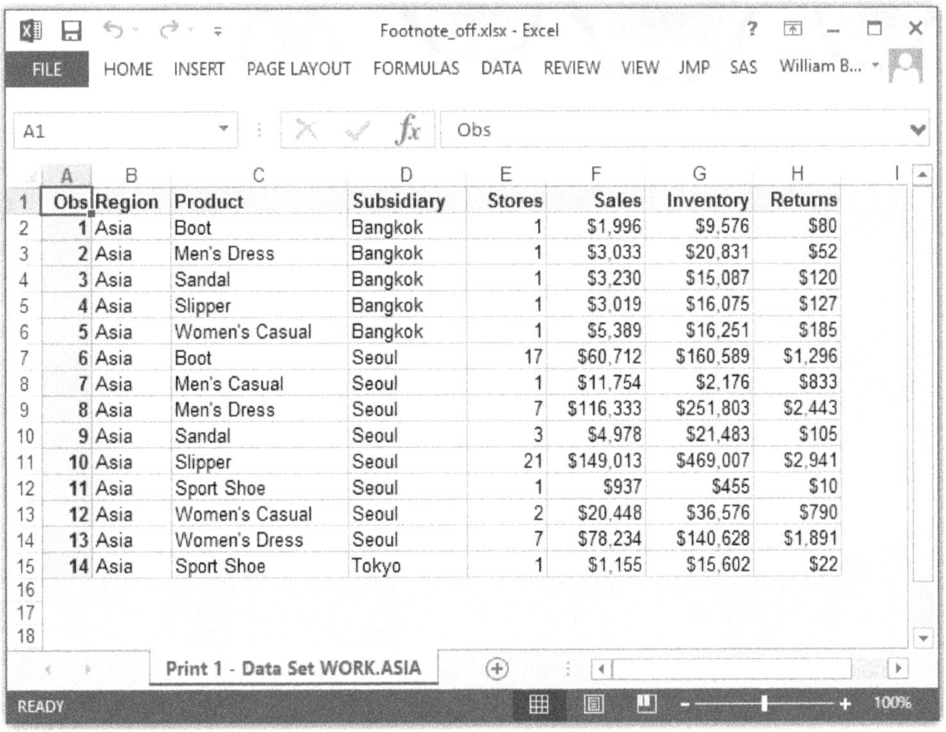
Footnote_off.xlsx - Excel

Obs	Region	Product	Subsidiary	Stores	Sales	Inventory	Returns
1	Asia	Boot	Bangkok	1	$1,996	$9,576	$80
2	Asia	Men's Dress	Bangkok	1	$3,033	$20,831	$52
3	Asia	Sandal	Bangkok	1	$3,230	$15,087	$120
4	Asia	Slipper	Bangkok	1	$3,019	$16,075	$127
5	Asia	Women's Casual	Bangkok	1	$5,389	$16,251	$185
6	Asia	Boot	Seoul	17	$60,712	$160,589	$1,296
7	Asia	Men's Casual	Seoul	1	$11,754	$2,176	$833
8	Asia	Men's Dress	Seoul	7	$116,333	$251,803	$2,443
9	Asia	Sandal	Seoul	3	$4,978	$21,483	$105
10	Asia	Slipper	Seoul	21	$149,013	$469,007	$2,941
11	Asia	Sport Shoe	Seoul	1	$937	$455	$10
12	Asia	Women's Casual	Seoul	2	$20,448	$36,576	$790
13	Asia	Women's Dress	Seoul	7	$78,234	$140,628	$1,891
14	Asia	Sport Shoe	Tokyo	1	$1,155	$15,602	$22

Print 1 - Data Set WORK.ASIA

With the EMBEDDED_FOOTNOTES= option turned off, no SAS footnotes appear at the bottom of the output worksheet. This is the default setting for the EMBEDDED_FOOTNOTES= suboption.This works well for most applications because it does not place alphabetic characters in any columns of the output worksheet.

Embedded Footnotes Turned On

When EMBEDDED_FOOTNOTES are turned on, then footnotes sent from SAS are displayed on the Excel worksheet pages. See SAS Code 7-2 for the code.

SAS Code 7-2 – Using the EMBEDDED_FOOTNOTES='ON' Option

```
** Code to generate the Excel file in Figure 7-2 ***;
ods excel file = "&path\Footnote_on.xlsx"
             options(EMBEDDED_FOOTNOTES='on');
   proc print data=Asia_only;
   run;
ods excel close;
```

Figure 7-2 – Excel Output with the EMBEDDED_FOOTNOTES= Option On

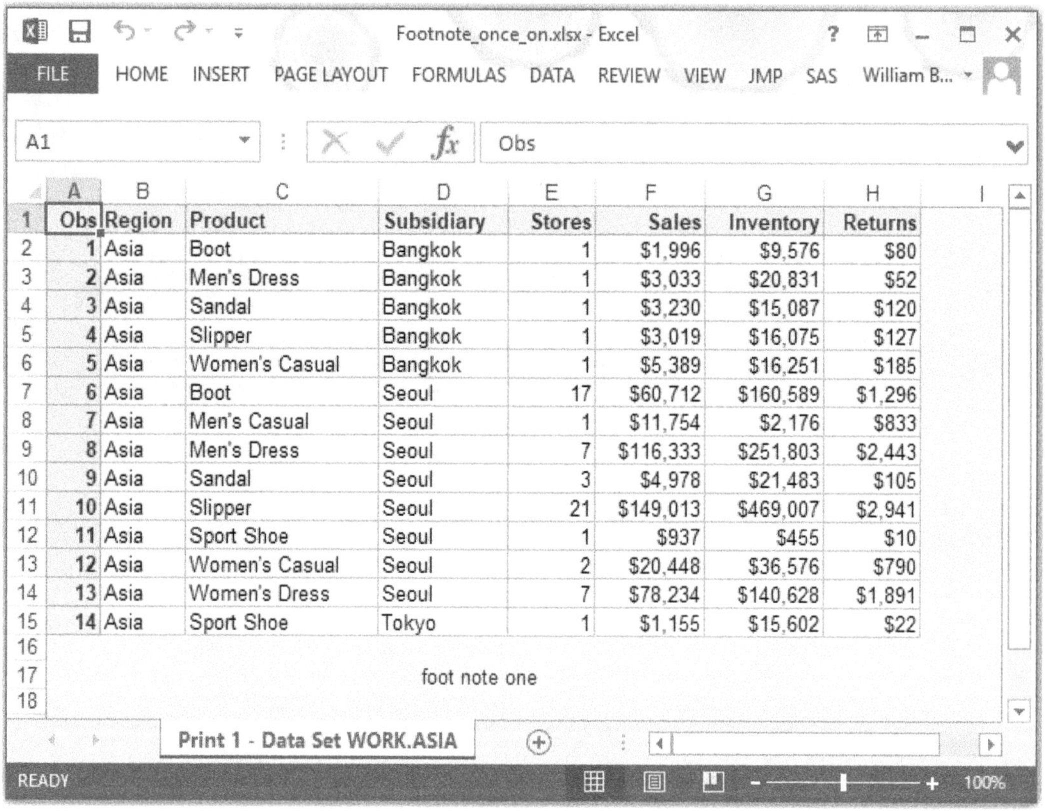

Obs	Region	Product	Subsidiary	Stores	Sales	Inventory	Returns
1	Asia	Boot	Bangkok	1	$1,996	$9,576	$80
2	Asia	Men's Dress	Bangkok	1	$3,033	$20,831	$52
3	Asia	Sandal	Bangkok	1	$3,230	$15,087	$120
4	Asia	Slipper	Bangkok	1	$3,019	$16,075	$127
5	Asia	Women's Casual	Bangkok	1	$5,389	$16,251	$185
6	Asia	Boot	Seoul	17	$60,712	$160,589	$1,296
7	Asia	Men's Casual	Seoul	1	$11,754	$2,176	$833
8	Asia	Men's Dress	Seoul	7	$116,333	$251,803	$2,443
9	Asia	Sandal	Seoul	3	$4,978	$21,483	$105
10	Asia	Slipper	Seoul	21	$149,013	$469,007	$2,941
11	Asia	Sport Shoe	Seoul	1	$937	$455	$10
12	Asia	Women's Casual	Seoul	2	$20,448	$36,576	$790
13	Asia	Women's Dress	Seoul	7	$78,234	$140,628	$1,891
14	Asia	Sport Shoe	Tokyo	1	$1,155	$15,602	$22

foot note one

Figure 7-2 shows a worksheet with the EMBEDDED_FOOTNOTES= option turned on. The SAS footnote appears at the bottom of the output worksheet. Here the suboption EMBEDDED_FOOTNOTES produces the footnote text 'footnote one' after the bottom of the output table.

Embedded Footnotes='ON' with Multiple Outputs on a Worksheet

When using the SHEET_INTERVAL='NONE' suboption with the EMBEDDED_FOOTNOTES= suboption, multiple SAS outputs can appear on the same worksheet and footnotes can be displayed for each output from SAS. See SAS Code 7-3 for a coding example and Figure 7-3 for an example of the Excel output.

SAS Code 7-3 – Using EMBEDDED_FOOTNOTES and SHEET_INTERVAL Suboptions.

```
** Code to generate the Excel file in Figure 7-3 ***;
ods excel file = "&path\Footnote_on_multiple.xlsx"
          options(EMBEDDED_FOOTNOTES='on'
                  SHEET_INTERVAL='none');
   proc print data=Asia_only;
   run;
   footnote 'footnote two';
   proc print data=Asia_only;
   run;
ods excel close;
```

This code produces an Excel worksheet with two output tables and a footnote after each table. See Figure 7-3 for details.

Figure 7-3 – EMBEDDED_FOOTNOTES= Option On and Multiple Outputs on One Worksheet

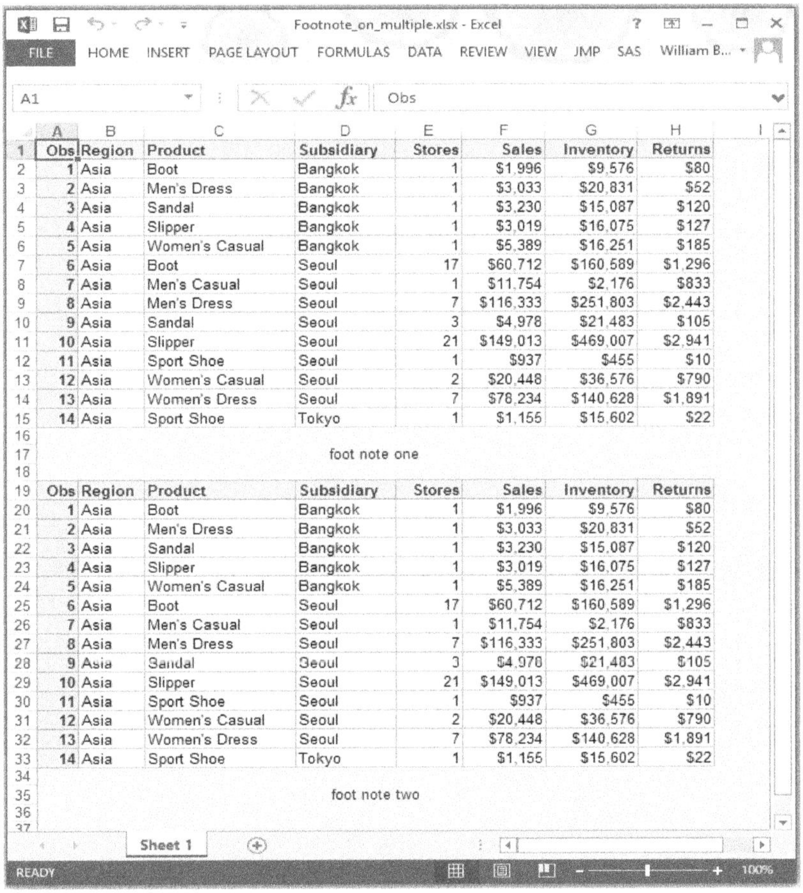

In Figure 7-3, the EMBEDDED_FOOTNOTES= option produces the footnote text 'foot note one' after the first output table and the footnote 'foot note two' after the second table. Both of these output tables were sent to the same worksheet with the SHEET_INTERVAL=Suboption.

The EMBED_FOOTNOTES_ONCE= Suboption

In SAS Code 7-4 the EMBED_FOOTNOTES_ONCE= suboption is used in conjunction with the EMBEDDED_FOOTNOTES= suboption. Because SAS Code 7-4 only outputs one SAS table to Excel, the EMBED_FOOTNOTES_ONCE suboption as it is used here looks like extra code that does nothing. However, the code in SAS Code 7-5 sends two tables to the output Excel worksheet. The output with EMBEDDED_FOOTNOTES='OFF' is not shown here because it looks the same as Figure 7-1.

EMBED_FOOTNOTES_ONCE with Only One Output Table

SAS Code 7-4 – Showing How to Print Footnotes Only Once on a Worksheet

```
Footnote 'foot note one';
ods excel file = "&path\Footnote_once_on.xlsx"
          options(EMBEDDED_FOOTNOTES='on'
                  EMBED_FOOTNOTES_ONCE='on');
   proc print data=Asia_only;
   run;
ods excel close;
```

Only one table and one footnote is output by the SAS Code 7-4 example.

Figure 7-4 – Excel Output with EMBEDDED_FOOTNOTES and EMBED_FOOTNOTES_ONCE Set to 'ON'

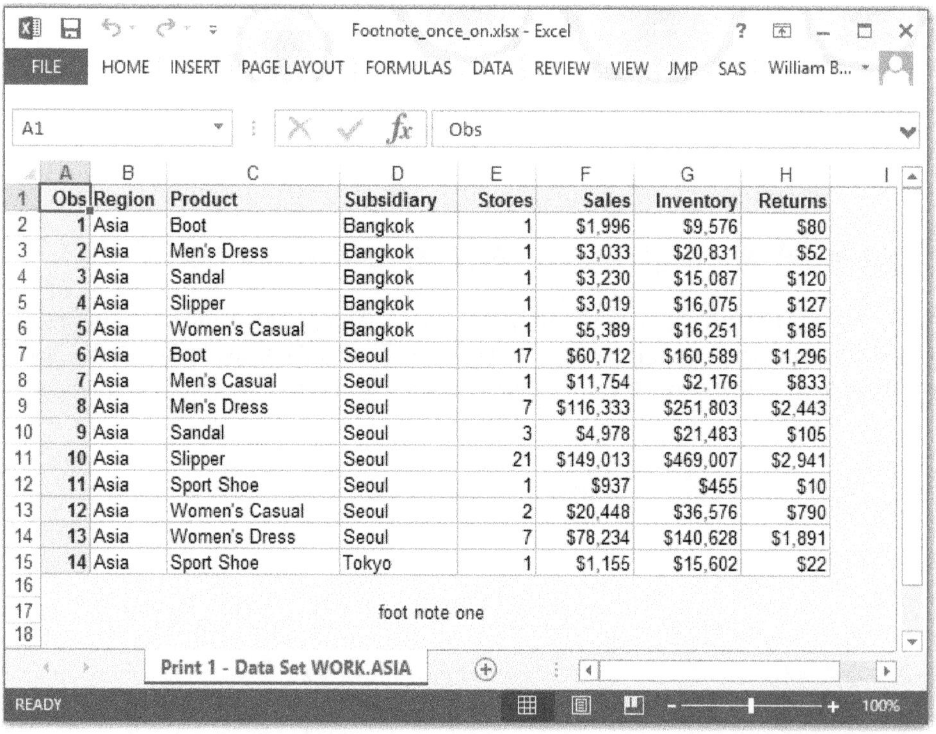

Obs	Region	Product	Subsidiary	Stores	Sales	Inventory	Returns
1	Asia	Boot	Bangkok	1	$1,996	$9,576	$80
2	Asia	Men's Dress	Bangkok	1	$3,033	$20,831	$52
3	Asia	Sandal	Bangkok	1	$3,230	$15,087	$120
4	Asia	Slipper	Bangkok	1	$3,019	$16,075	$127
5	Asia	Women's Casual	Bangkok	1	$5,389	$16,251	$185
6	Asia	Boot	Seoul	17	$60,712	$160,589	$1,296
7	Asia	Men's Casual	Seoul	1	$11,754	$2,176	$833
8	Asia	Men's Dress	Seoul	7	$116,333	$251,803	$2,443
9	Asia	Sandal	Seoul	3	$4,978	$21,483	$105
10	Asia	Slipper	Seoul	21	$149,013	$469,007	$2,941
11	Asia	Sport Shoe	Seoul	1	$937	$455	$10
12	Asia	Women's Casual	Seoul	2	$20,448	$36,576	$790
13	Asia	Women's Dress	Seoul	7	$78,234	$140,628	$1,891
14	Asia	Sport Shoe	Tokyo	1	$1,155	$15,602	$22

It looks like the SAS Code 7-4 did nothing, but look at SAS Code 7-5 to see a different result.

EMBED_FOOTNOTES_ONCE with More Than One Output Table

In SAS Code 7-5 two footnotes and two output tables are coded and the result is shown in Figure 7-5. The EMBED_FOOTNOTES_ONCE= suboption forced only the last footnote to appear at the bottom of each page. However, footnotes can be changed if the output Excel worksheet changes.

SAS Code 7-5 – Printing the Last Footnote Only Once on a Worksheet

```
Footnote 'foot note one';
ods excel file = "&path\Footnote_once_on_multiple.xlsx"
          options(EMBEDDED_FOOTNOTES='on'
                  EMBED_FOOTNOTES_ONCE='on'
                  SHEET_INTERVAL='none');
   proc print data=Asia_only;
   run;
   Footnote 'foot note two';
   proc print data=Asia_only;
   run;
ods excel close;
```

The footnote 'foot note one' will not appear on the output Excel worksheet because it is not the last footnote output to the worksheet. SHEET_INTERVAL='NONE' causes both tables to appear on the same Excel worksheet.

Figure 7-5 – Excel Output with EMBED_FOOTNOTES_ONCE and Multiple Outputs on a Page

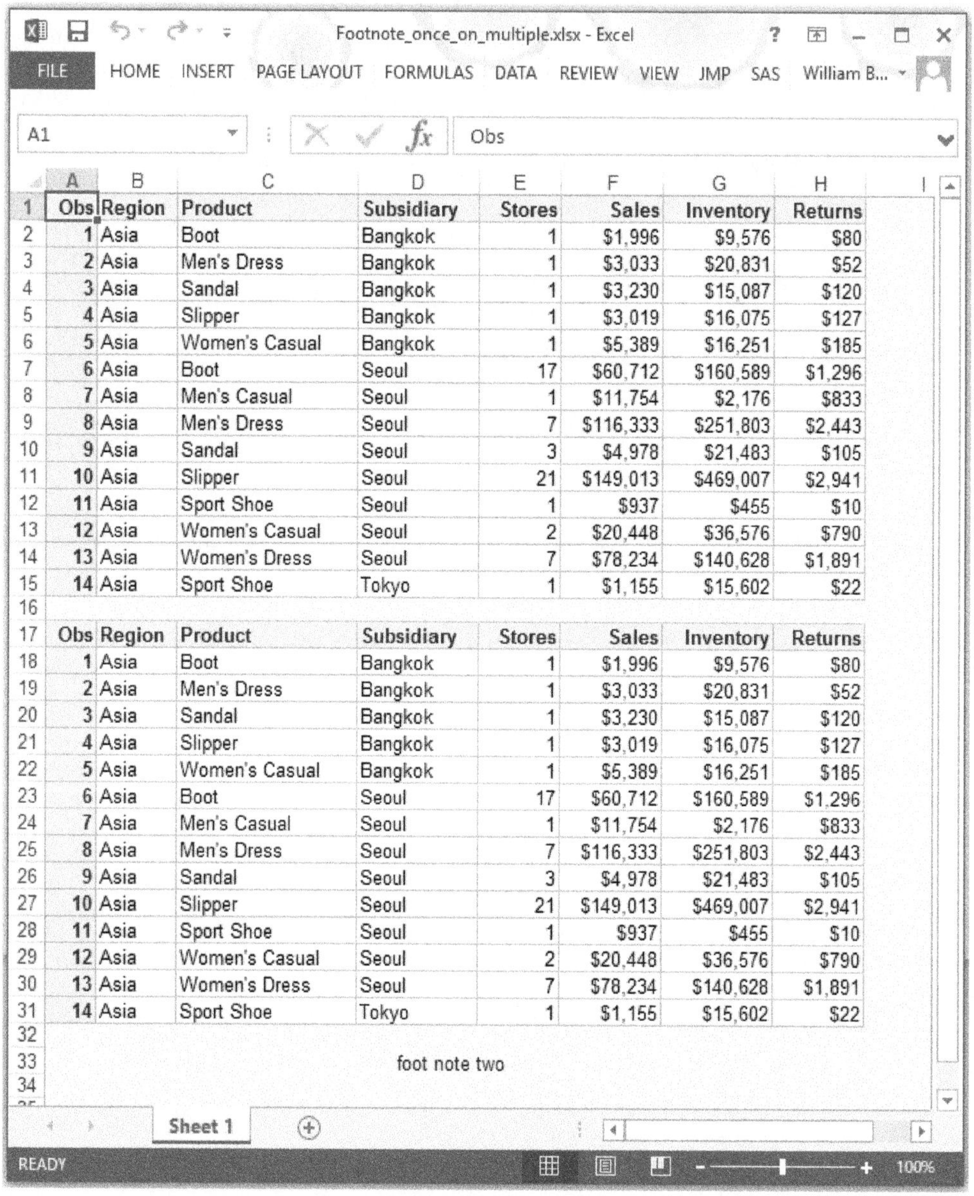

	A	B	C	D	E	F	G	H
1	Obs	Region	Product	Subsidiary	Stores	Sales	Inventory	Returns
2	1	Asia	Boot	Bangkok	1	$1,996	$9,576	$80
3	2	Asia	Men's Dress	Bangkok	1	$3,033	$20,831	$52
4	3	Asia	Sandal	Bangkok	1	$3,230	$15,087	$120
5	4	Asia	Slipper	Bangkok	1	$3,019	$16,075	$127
6	5	Asia	Women's Casual	Bangkok	1	$5,389	$16,251	$185
7	6	Asia	Boot	Seoul	17	$60,712	$160,589	$1,296
8	7	Asia	Men's Casual	Seoul	1	$11,754	$2,176	$833
9	8	Asia	Men's Dress	Seoul	7	$116,333	$251,803	$2,443
10	9	Asia	Sandal	Seoul	3	$4,978	$21,483	$105
11	10	Asia	Slipper	Seoul	21	$149,013	$469,007	$2,941
12	11	Asia	Sport Shoe	Seoul	1	$937	$455	$10
13	12	Asia	Women's Casual	Seoul	2	$20,448	$36,576	$790
14	13	Asia	Women's Dress	Seoul	7	$78,234	$140,628	$1,891
15	14	Asia	Sport Shoe	Tokyo	1	$1,155	$15,602	$22
16								
17	Obs	Region	Product	Subsidiary	Stores	Sales	Inventory	Returns
18	1	Asia	Boot	Bangkok	1	$1,996	$9,576	$80
19	2	Asia	Men's Dress	Bangkok	1	$3,033	$20,831	$52
20	3	Asia	Sandal	Bangkok	1	$3,230	$15,087	$120
21	4	Asia	Slipper	Bangkok	1	$3,019	$16,075	$127
22	5	Asia	Women's Casual	Bangkok	1	$5,389	$16,251	$185
23	6	Asia	Boot	Seoul	17	$60,712	$160,589	$1,296
24	7	Asia	Men's Casual	Seoul	1	$11,754	$2,176	$833
25	8	Asia	Men's Dress	Seoul	7	$116,333	$251,803	$2,443
26	9	Asia	Sandal	Seoul	3	$4,978	$21,483	$105
27	10	Asia	Slipper	Seoul	21	$149,013	$469,007	$2,941
28	11	Asia	Sport Shoe	Seoul	1	$937	$455	$10
29	12	Asia	Women's Casual	Seoul	2	$20,448	$36,576	$790
30	13	Asia	Women's Dress	Seoul	7	$78,234	$140,628	$1,891
31	14	Asia	Sport Shoe	Tokyo	1	$1,155	$15,602	$22
32								
33				foot note two				
34								

Here the combination of EMBED_FOOTNOTES_ONCE and EMBEDDED_FOOTNOTES suppresses the footnote 'foot note one' that was displayed in Figure 7-3. Only the most active footnote is displayed. This is often used with the EMBEDDED_TITLES option to place the SAS titles and footnotes onto the Excel worksheet pages. When multiple pages are produced the footnotes appear on each page,

The EMBEDDED_TITLES= Suboption

The EMBEDDED_TITLES= suboption shown here sends the SAS titles to the output Excel worksheet. This is often used with the EMBEDDED_FOOTNOTES= suboption to place the SAS titles and footnotes onto the Excel worksheet pages. When multiple pages are produced the titles appear on each page. The titles can be changed and the new title will appear on the next output worksheet. When multiple outputs are sent to one worksheet, the titles can be changed between procedure outputs on the same worksheet.

EMBEDDED_TITLES='OFF'

EMBBEDDED_TITLES ='OFF' is the default value for this suboption. SAS Code 7-6 shows how to code this suboption. Because you can code many ODS EXCEL statements, you can turn this suboption on and off, but turning it on and off is only effective if the worksheet changes.

SAS Code 7-6 – Using the EMBEDDED_TITLES='OFF' Suboption

```
Title 'title one';
ods excel file = "&path\Title_off.xlsx"
          options(EMBEDDED_TITLES='off');
   proc print data=Asia_only;
   run;
ods excel close;
```

The SAS Code 7-6 above is the same as the default and it produces a worksheet with no titles, as shown in Figure 7-6 below.

Figure 7-6 – Excel Output with EMBEDDED_TITLES='OFF'

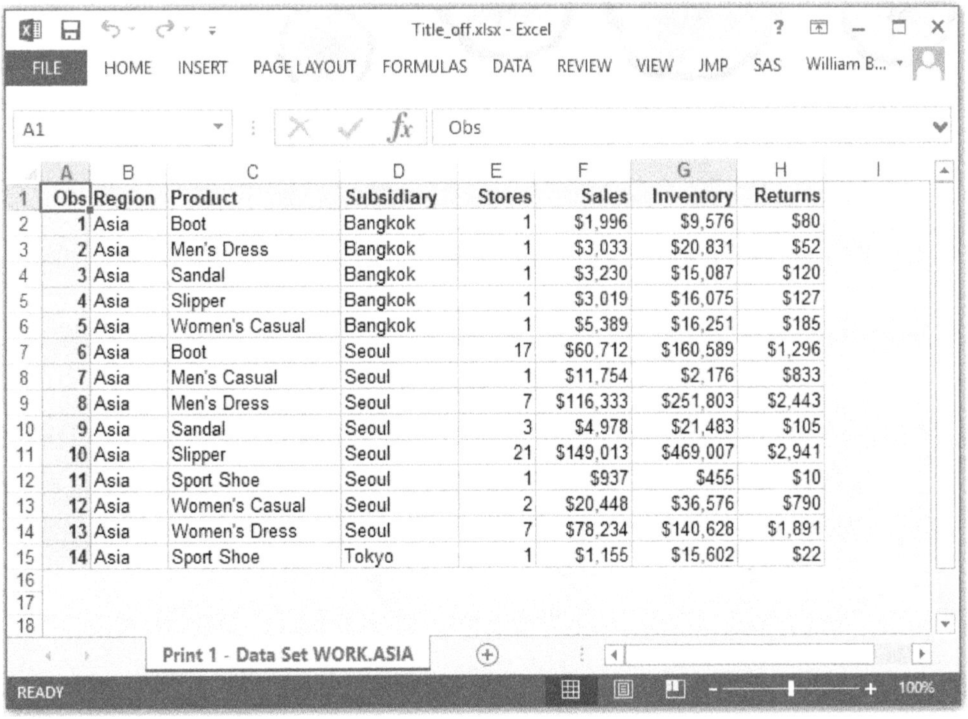

Obs	Region	Product	Subsidiary	Stores	Sales	Inventory	Returns
1	Asia	Boot	Bangkok	1	$1,996	$9,576	$80
2	Asia	Men's Dress	Bangkok	1	$3,033	$20,831	$52
3	Asia	Sandal	Bangkok	1	$3,230	$15,087	$120
4	Asia	Slipper	Bangkok	1	$3,019	$16,075	$127
5	Asia	Women's Casual	Bangkok	1	$5,389	$16,251	$185
6	Asia	Boot	Seoul	17	$60,712	$160,589	$1,296
7	Asia	Men's Casual	Seoul	1	$11,754	$2,176	$833
8	Asia	Men's Dress	Seoul	7	$116,333	$251,803	$2,443
9	Asia	Sandal	Seoul	3	$4,978	$21,483	$105
10	Asia	Slipper	Seoul	21	$149,013	$469,007	$2,941
11	Asia	Sport Shoe	Seoul	1	$937	$455	$10
12	Asia	Women's Casual	Seoul	2	$20,448	$36,576	$790
13	Asia	Women's Dress	Seoul	7	$78,234	$140,628	$1,891
14	Asia	Sport Shoe	Tokyo	1	$1,155	$15,602	$22

This is the default output when the EMBEDDED_TITLES= option is either set to 'OFF' or not used.

EMBEDDED_TITLES='ON'

The EMBEDDED_TITLES='ON' suboption causes the SAS TITLE to be output in the Excel worksheets that are generated by the ODS Excel output. Figure 7-7 shows the Excel output worksheet created by SAS Code 7-7.

SAS Code 7-7 – Using EMBEDDED_TITLES='ON'

```
Title 'title one';

ods excel file = "&path\Title_on.xlsx"
          options(EMBEDDED_TITLES='on');
   proc print data=Asia_only;
   run;
ods excel close;
```

Figure 7-7 – Excel Output with EMBEDDED_TITLES='ON'

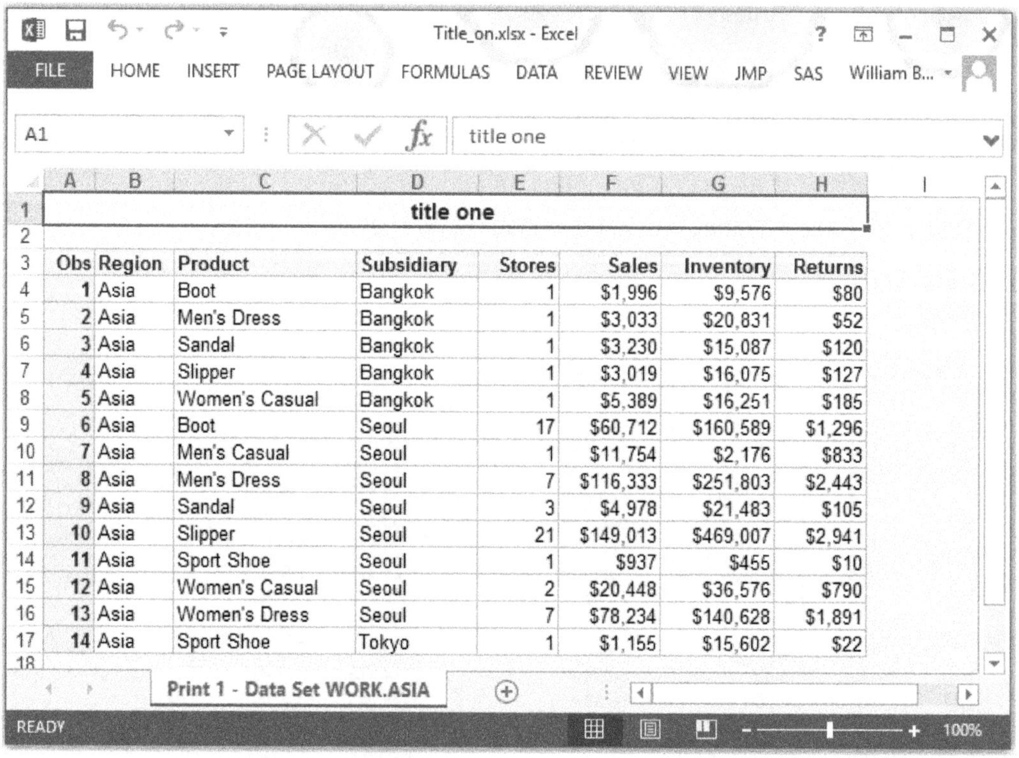

Figure 7-7 shows the title at the top of the Excel worksheet.

SAS Code 7-8 – EMBEDDED_TITLES='ON' and SHEET_INTERVAL='NONE'

```
Title 'title one';

ods excel file = "&path\Titles_on_multiple.xlsx"
           options(EMBEDDED_TITLES='on'
                   SHEET_INTERVAL='none');
   proc print data=Asia_only;
   run;
   Title 'title two';
   proc print data=Asia_only;
   run;
ods excel close;
```

SAS Code 7-8 produces the output in Figure 7-8 showing two output tables each with its own title.

Figure 7-8 – Excel Output with EMBEDDED_TITLES='ON' and Two SAS TITLE Statements

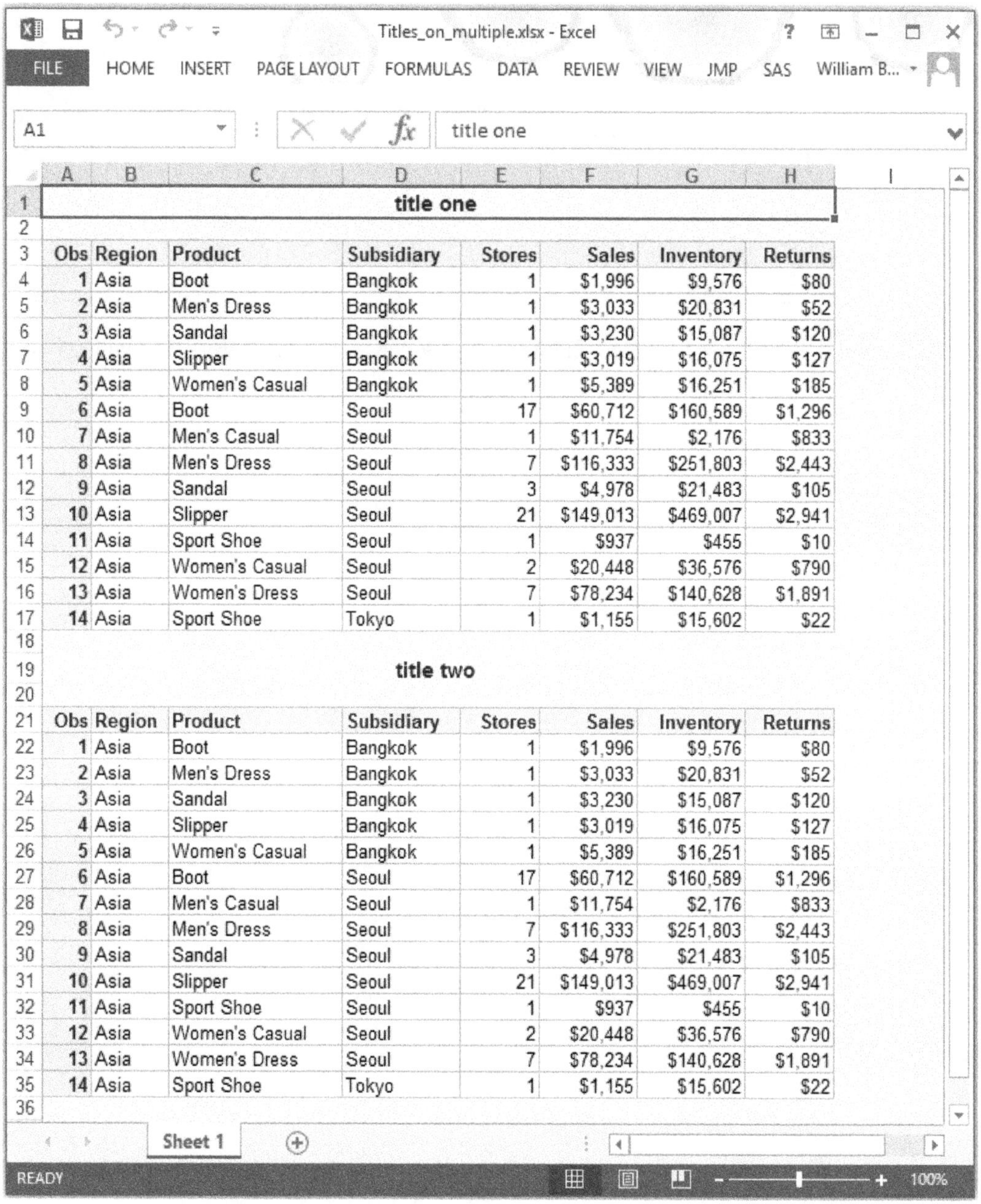

title one

Obs	Region	Product	Subsidiary	Stores	Sales	Inventory	Returns
1	Asia	Boot	Bangkok	1	$1,996	$9,576	$80
2	Asia	Men's Dress	Bangkok	1	$3,033	$20,831	$52
3	Asia	Sandal	Bangkok	1	$3,230	$15,087	$120
4	Asia	Slipper	Bangkok	1	$3,019	$16,075	$127
5	Asia	Women's Casual	Bangkok	1	$5,389	$16,251	$185
6	Asia	Boot	Seoul	17	$60,712	$160,589	$1,296
7	Asia	Men's Casual	Seoul	1	$11,754	$2,176	$833
8	Asia	Men's Dress	Seoul	7	$116,333	$251,803	$2,443
9	Asia	Sandal	Seoul	3	$4,978	$21,483	$105
10	Asia	Slipper	Seoul	21	$149,013	$469,007	$2,941
11	Asia	Sport Shoe	Seoul	1	$937	$455	$10
12	Asia	Women's Casual	Seoul	2	$20,448	$36,576	$790
13	Asia	Women's Dress	Seoul	7	$78,234	$140,628	$1,891
14	Asia	Sport Shoe	Tokyo	1	$1,155	$15,602	$22

title two

Obs	Region	Product	Subsidiary	Stores	Sales	Inventory	Returns
1	Asia	Boot	Bangkok	1	$1,996	$9,576	$80
2	Asia	Men's Dress	Bangkok	1	$3,033	$20,831	$52
3	Asia	Sandal	Bangkok	1	$3,230	$15,087	$120
4	Asia	Slipper	Bangkok	1	$3,019	$16,075	$127
5	Asia	Women's Casual	Bangkok	1	$5,389	$16,251	$185
6	Asia	Boot	Seoul	17	$60,712	$160,589	$1,296
7	Asia	Men's Casual	Seoul	1	$11,754	$2,176	$833
8	Asia	Men's Dress	Seoul	7	$116,333	$251,803	$2,443
9	Asia	Sandal	Seoul	3	$4,978	$21,483	$105
10	Asia	Slipper	Seoul	21	$149,013	$469,007	$2,941
11	Asia	Sport Shoe	Seoul	1	$937	$455	$10
12	Asia	Women's Casual	Seoul	2	$20,448	$36,576	$790
13	Asia	Women's Dress	Seoul	7	$78,234	$140,628	$1,891
14	Asia	Sport Shoe	Tokyo	1	$1,155	$15,602	$22

Figure 7-8 shows how the titles can be changed between procedure output tables. Here the SHEET_INTERVAL= option 'NONE' was used to force the SAS output from two procedures onto one Excel worksheet.

The EMBED_TITLES_ONCE= Suboption

The EMBED_TITLES_ONCE= suboption shown here works with the EMBEDDED_TITLES= suboption. The SAS titles are sent to the output Excel worksheet with the EMBEDDED_TITLES= option. EMBED_TITLES_ONCE makes sure that the titles appear only once at the top of the page.

EMBED_TITLES_ONCE and Only One Output Worksheet

The EMBED_TITLES_ONCE= suboption places the first active SAS title onto the Excel worksheet pages. When multiple pages are produced the titles appear on each page. The titles can be changed and the new title will appear on the next output worksheet. When multiple outputs are sent to one worksheet the footnotes can be changed between outputs, but only the first active set of titles will be displayed.

SAS Code 7-9 – The EMBED_TITLES_ONCE= Suboption

```
Title 'title one';
ods excel file = "&path\Title_once_on.xlsx"
          options(EMBEDDED_TITLES='on'
                  EMBED_TITLES_ONCE='on');
   proc print data=Asia_only;
   run;
ods excel close;
```

When only one TITLE statement is active it is used.

Figure 7-9 – EMBED_TITLES_ONCE= Suboption Set to 'ON'

Obs	Region	Product	Subsidiary	Stores	Sales	Inventory	Returns
1	Asia	Boot	Bangkok	1	$1,996	$9,576	$80
2	Asia	Men's Dress	Bangkok	1	$3,033	$20,831	$52
3	Asia	Sandal	Bangkok	1	$3,230	$15,087	$120
4	Asia	Slipper	Bangkok	1	$3,019	$16,075	$127
5	Asia	Women's Casual	Bangkok	1	$5,389	$16,251	$185
6	Asia	Boot	Seoul	17	$60,712	$160,589	$1,296
7	Asia	Men's Casual	Seoul	1	$11,754	$2,176	$833
8	Asia	Men's Dress	Seoul	7	$116,333	$251,803	$2,443
9	Asia	Sandal	Seoul	3	$4,978	$21,483	$105
10	Asia	Slipper	Seoul	21	$149,013	$469,007	$2,941
11	Asia	Sport Shoe	Seoul	1	$937	$455	$10
12	Asia	Women's Casual	Seoul	2	$20,448	$36,576	$790
13	Asia	Women's Dress	Seoul	7	$78,234	$140,628	$1,891
14	Asia	Sport Shoe	Tokyo	1	$1,155	$15,602	$22

Note that when only one table is output and if both the EMBEDDED_TITLES= and EMBED_TITLES_ONCE= suboptions are used, then the output appears to look the same.

EMBED_TITLES_ONCE with Multiple Outputs on a Worksheet

When writing more than one SAS output to the same Excel worksheet, EMBEDDED_TITLES='ON' and EMBED_TITLES_ONCE='ON' work together to display the SAS titles in the Excel worksheet. If two titles would normally appear on the worksheet, the EMBED_TITLES_ONCE= option allows only the first active title to be output.

SAS Code 7-10 – EMBED_TITLES_ONCE Used with Two Titles and Output Tables

```
Title 'title one';

ods excel file = "&path\Title_once_on_multiple.xlsx"
          options(EMBEDDED_TITLES='on'
                  EMBED_TITLES_ONCE='on'
                  SHEET_INTERVAL='none');
   proc print data=Asia_only;
   run;
   Title 'title two';
   proc print data=Asia_only;
   run;
ods excel close;
```

Figure 7-10 – EMBEDDED_TITLES and EMBED_TITLES_ONCE Suboptions Used with Multiple Titles

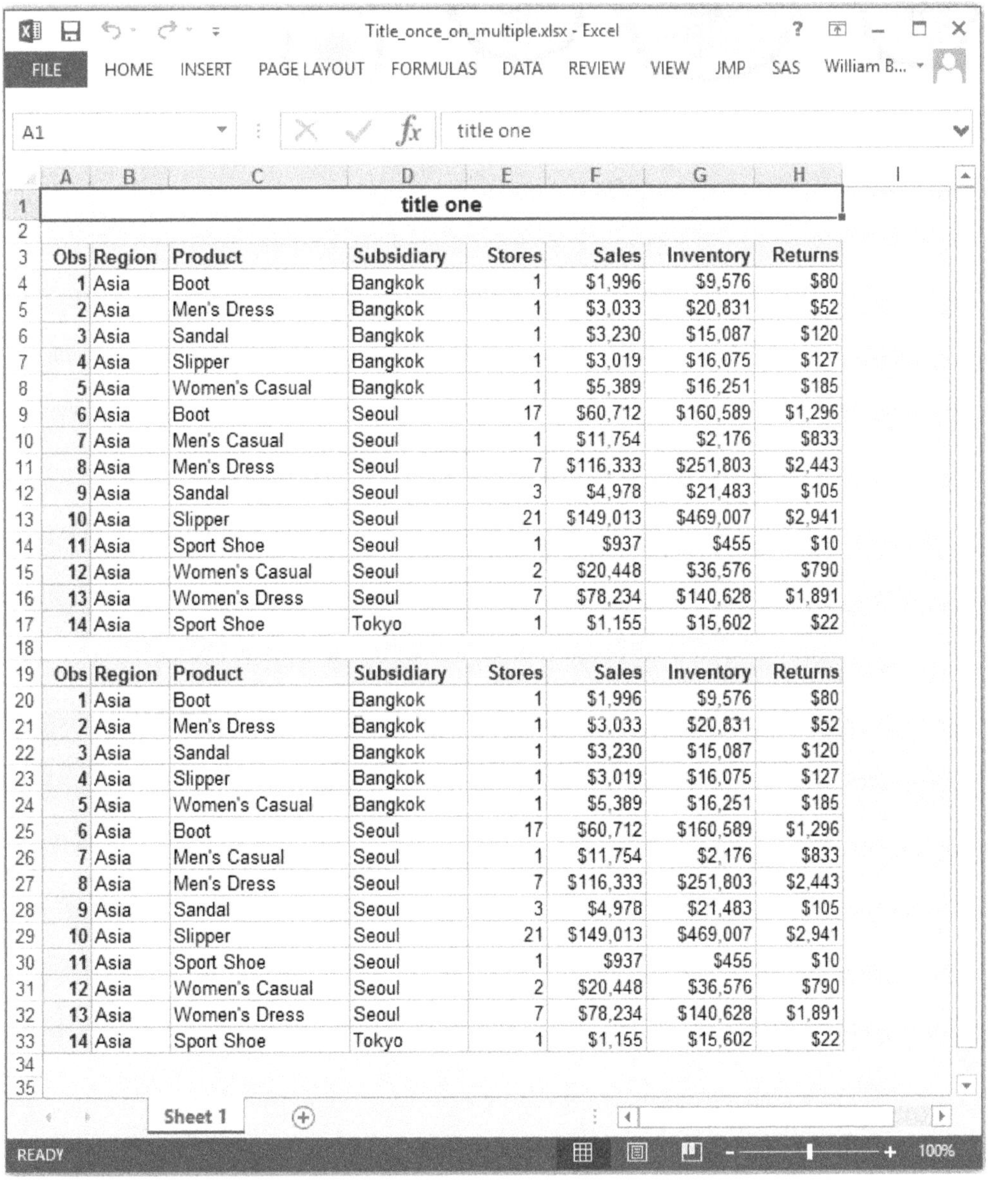

Obs	Region	Product	Subsidiary	Stores	Sales	Inventory	Returns
			title one				
Obs	Region	Product	Subsidiary	Stores	Sales	Inventory	Returns
1	Asia	Boot	Bangkok	1	$1,996	$9,576	$80
2	Asia	Men's Dress	Bangkok	1	$3,033	$20,831	$52
3	Asia	Sandal	Bangkok	1	$3,230	$15,087	$120
4	Asia	Slipper	Bangkok	1	$3,019	$16,075	$127
5	Asia	Women's Casual	Bangkok	1	$5,389	$16,251	$185
6	Asia	Boot	Seoul	17	$60,712	$160,589	$1,296
7	Asia	Men's Casual	Seoul	1	$11,754	$2,176	$833
8	Asia	Men's Dress	Seoul	7	$116,333	$251,803	$2,443
9	Asia	Sandal	Seoul	3	$4,978	$21,483	$105
10	Asia	Slipper	Seoul	21	$149,013	$469,007	$2,941
11	Asia	Sport Shoe	Seoul	1	$937	$455	$10
12	Asia	Women's Casual	Seoul	2	$20,448	$36,576	$790
13	Asia	Women's Dress	Seoul	7	$78,234	$140,628	$1,891
14	Asia	Sport Shoe	Tokyo	1	$1,155	$15,602	$22
Obs	Region	Product	Subsidiary	Stores	Sales	Inventory	Returns
1	Asia	Boot	Bangkok	1	$1,996	$9,576	$80
2	Asia	Men's Dress	Bangkok	1	$3,033	$20,831	$52
3	Asia	Sandal	Bangkok	1	$3,230	$15,087	$120
4	Asia	Slipper	Bangkok	1	$3,019	$16,075	$127
5	Asia	Women's Casual	Bangkok	1	$5,389	$16,251	$185
6	Asia	Boot	Seoul	17	$60,712	$160,589	$1,296
7	Asia	Men's Casual	Seoul	1	$11,754	$2,176	$833
8	Asia	Men's Dress	Seoul	7	$116,333	$251,803	$2,443
9	Asia	Sandal	Seoul	3	$4,978	$21,483	$105
10	Asia	Slipper	Seoul	21	$149,013	$469,007	$2,941
11	Asia	Sport Shoe	Seoul	1	$937	$455	$10
12	Asia	Women's Casual	Seoul	2	$20,448	$36,576	$790
13	Asia	Women's Dress	Seoul	7	$78,234	$140,628	$1,891
14	Asia	Sport Shoe	Tokyo	1	$1,155	$15,602	$22

Figure 7-10 shows how the titles can be changed between procedure output tables. Here the SHEET_INTERVAL=NONE option was used to force the SAS output from two procedures onto one Excel worksheet. EMBED_TITLES_ONCE='ON' suppresses the second title, but only the first active title is placed in the worksheet.

The FITTOPAGE= Suboption

The SAS ODS EXCEL suboption FITTOPAGE= is a handy way to send output directly to Excel to be printed on one page. However, Excel is in control of the printing and it will at times compress data into a print font that becomes unreadably small. It is up to the SAS programmer to make sure that the output being sent to Excel will be readable on one page. My example here deliberately fits that bill.

SAS Code 7-11 – The FITTOPAGE= Suboption

```
ods excel file = "&path\Fit_to_Page_off.xlsx"
          options(FITTOPAGE='off');
   proc print data=Asia_only;
   run;
ods excel close;
```

To determine whether the option worked is to look at the EXCEL Page Setup sheet shown in Figure 7-11. The "FIT TO" option on the Excel "Page Setup" sheet is not set.

Figure 7-11 – The FITTOPAGE= Default Suboption

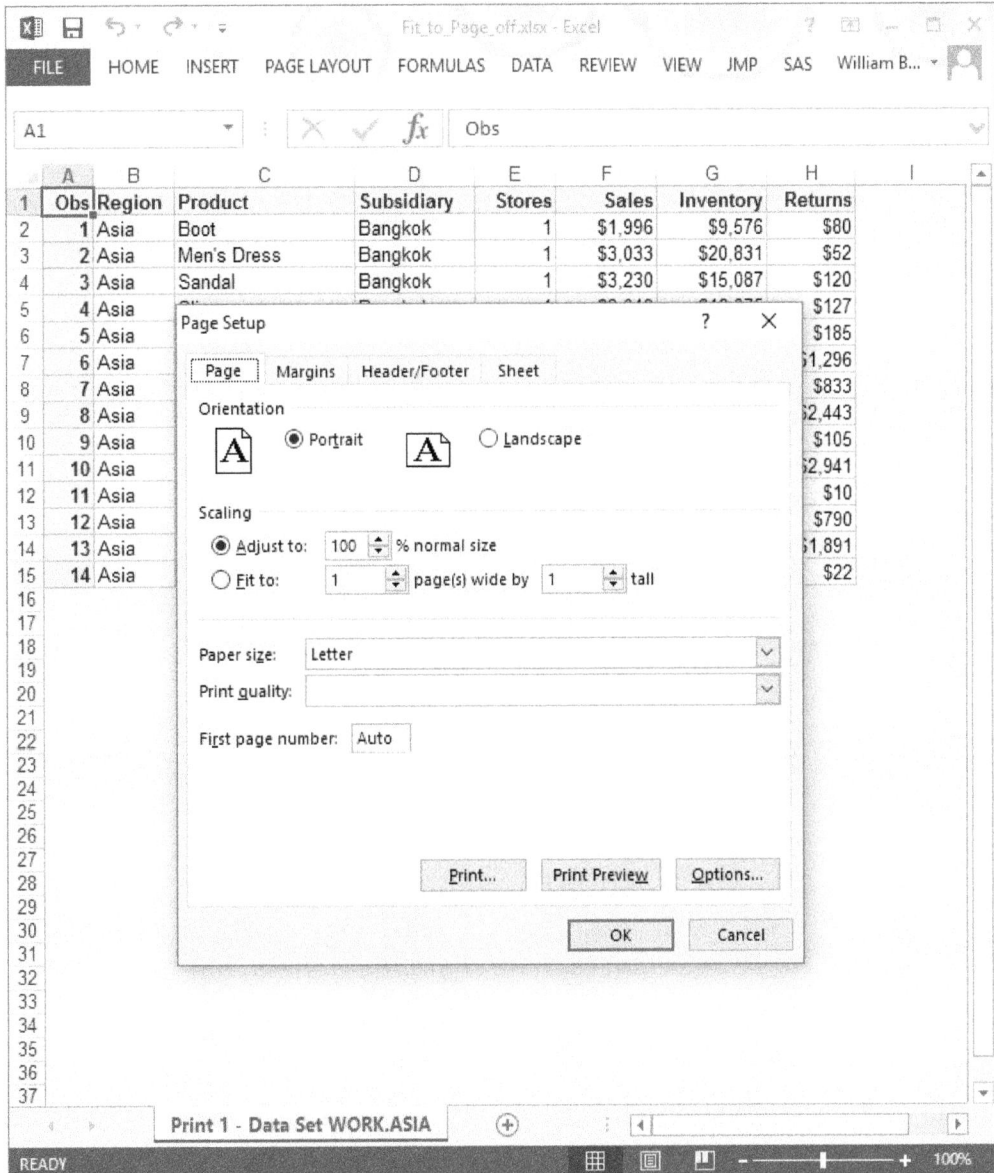

The image in Figure 7-11 is the default setting for Excel, which is the same as when the SAS suboption FITTOPAGE= is set to OFF. Therefore, the setting does not activate the "FIT TO" option within Excel under the "SCALING" settings. Consequently, the printed output will not be forced onto one page.

SAS Code 7-12 – The FITTOPAGE Suboption Set to 'ON'

```
ods excel file = "&path\Fit_to_Page_on.xlsx"
          options(FITTOPAGE='on');
   proc print data=Asia_only;
   run;
ods excel close;
```

Figure 7-12 – The FITTOPAGE Suboption Using the 'ON' Setting

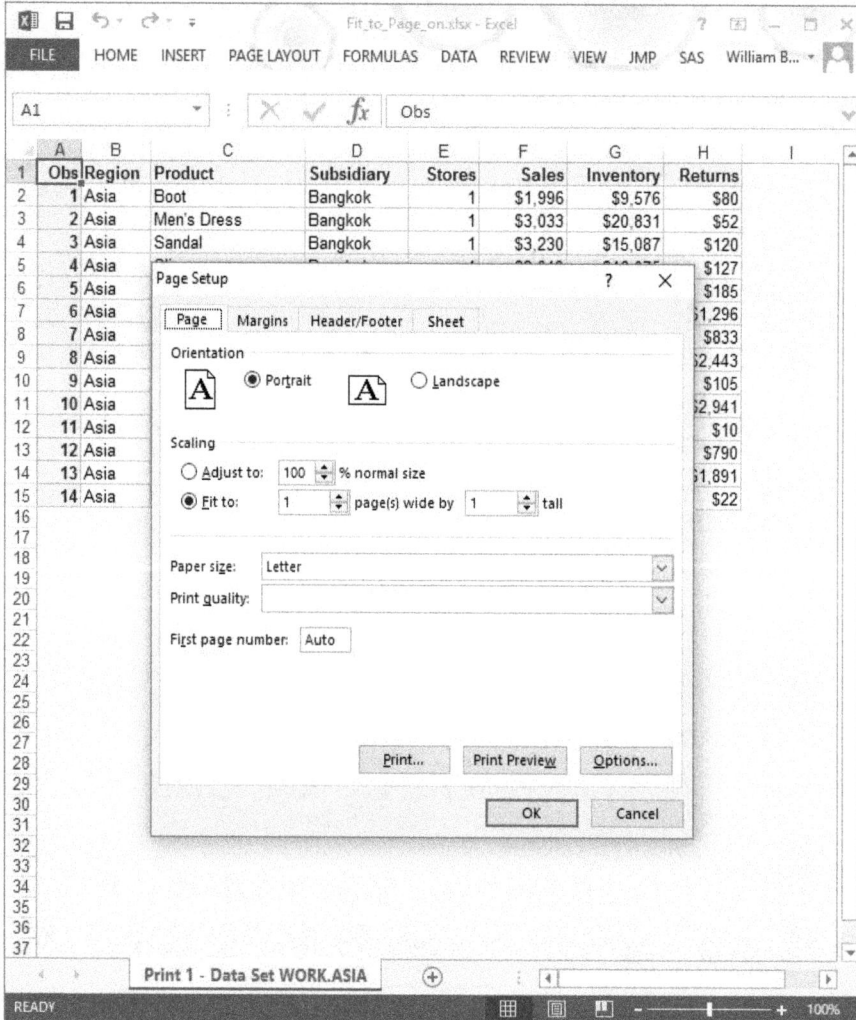

The image in Figure 7-12 activates the "FIT TO" suboption within Excel under the "SCALING" settings. The FITTOPAGE value used here is 'ON'.

The GRIDLINES= Suboption

Turning grid lines on and off in Excel usually requires accessing the Excel "Page Setup" (see the chapter introduction) to toggle a check box for the feature. I looked close to find the differences that the setting activa ted. It was only visible on the Excel print preview panel. The difference is a darker exterior grid line on the "Print Preview" panel when gridlines are on. In fact, when viewing the "Print Preview" panel for the GRIDLINES ='ON' output, there also appears to be a blank row at the bottom of the output tables.

SAS Code 7-13 – The GRIDLINES='OFF' Suboption

```
ods excel file = "&path\GRIDLINES_off.xlsx"
          options(GRIDLINES='off');
  proc print data=Asia_only;
  run;
ods excel close;
```

Figure 7-13 – Page Setup When GRIDLINES are turned 'OFF'

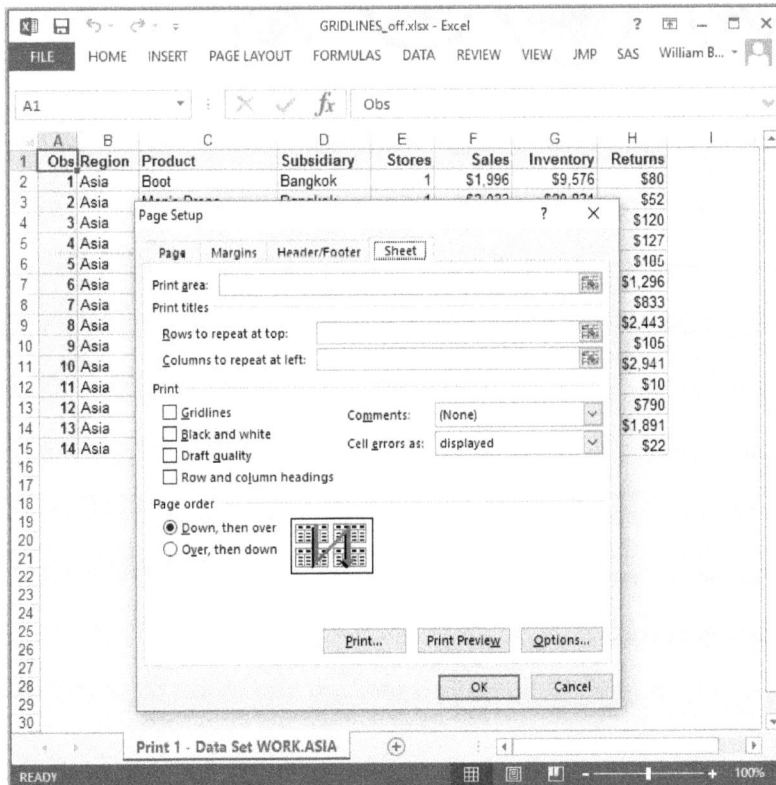

The GRIDLINES check box is blank in the "PRINT" section.

Figure 7-14 – "Print Preview" Sheet When GRIDLINES='OFF'

The SAS System

Obs	Region	Product	Subsidiary	Stores	Sales	Inventory	Returns
1	Asia	Boot	Bangkok	1	$1,996	$9,576	$80
2	Asia	Men's Dress	Bangkok	1	$3,033	$20,831	$52
3	Asia	Sandal	Bangkok	1	$3,230	$15,087	$120
4	Asia	Slipper	Bangkok	1	$3,019	$16,075	$127
5	Asia	Women's Casual	Bangkok	1	$5,389	$16,251	$185
6	Asia	Boot	Seoul	17	$60,712	$160,589	$1,296
7	Asia	Men's Casual	Seoul	1	$11,754	$2,176	$833
8	Asia	Men's Dress	Seoul	7	$116,333	$251,803	$2,443
9	Asia	Sandal	Seoul	3	$4,978	$21,483	$105
10	Asia	Slipper	Seoul	21	$149,013	$469,007	$2,941
11	Asia	Sport Shoe	Seoul	1	$937	$455	$10
12	Asia	Women's Casual	Seoul	2	$20,448	$36,576	$790
13	Asia	Women's Dress	Seoul	7	$78,234	$140,628	$1,891
14	Asia	Sport Shoe	Tokyo	1	$1,155	$15,602	$22

Notice that the exterior line around the data is the same as the interior lines of the data table. This indicates that GRIDLINES are turned 'OFF'.

SAS Code 7-14 – The GRIDLINES='ON' Suboption

```
ods excel file = "&path\GRIDLINES_on.xlsx"
          options(GRIDLINES='on');
   proc print data=Asia_only;
   run;
ods excel close;
```

Figure 7-15 – Page Setup When GRIDLINES Are Turned 'ON'

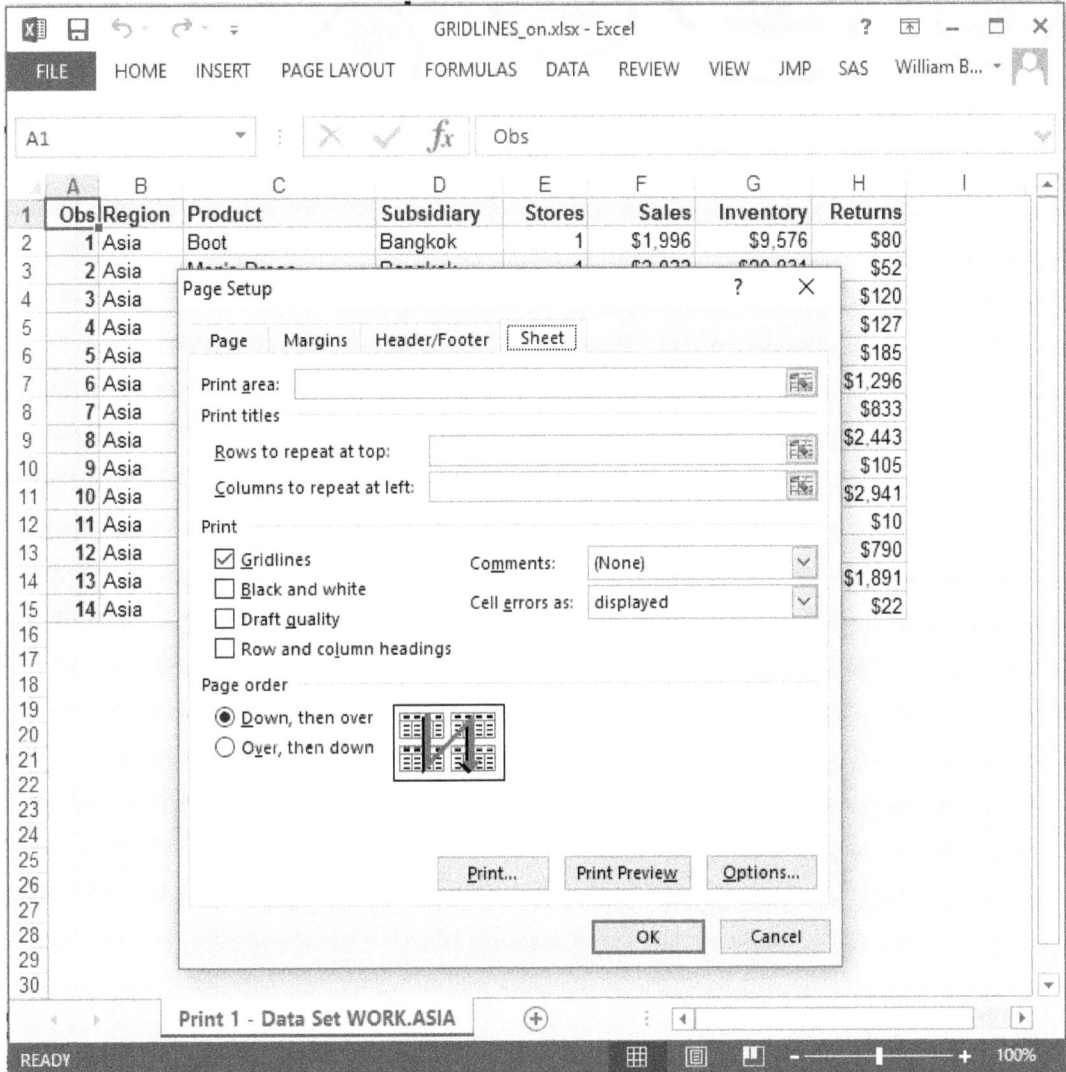

The GRIDLINES check box is selected in the "PRINT" section.

Figure 7-16 – "Print Preview" Sheet When GRIDLINES='ON'

The SAS System

Obs	Region	Product	Subsidiary	Stores	Sales	Inventory	Returns
1	Asia	Boot	Bangkok	1	$1,996	$9,576	$80
2	Asia	Men's Dress	Bangkok	1	$3,033	$20,831	$52
3	Asia	Sandal	Bangkok	1	$3,230	$15,087	$120
4	Asia	Slipper	Bangkok	1	$3,019	$16,075	$127
5	Asia	Women's Casual	Bangkok	1	$5,389	$16,251	$185
6	Asia	Boot	Seoul	17	$60,712	$160,589	$1,296
7	Asia	Men's Casual	Seoul	1	$11,754	$2,176	$833
8	Asia	Men's Dress	Seoul	7	$116,333	$251,803	$2,443
9	Asia	Sandal	Seoul	3	$4,978	$21,483	$105
10	Asia	Slipper	Seoul	21	$149,013	$469,007	$2,941
11	Asia	Sport Shoe	Seoul	1	$937	$455	$10
12	Asia	Women's Casual	Seoul	2	$20,448	$36,576	$790
13	Asia	Women's Dress	Seoul	7	$78,234	$140,628	$1,891
14	Asia	Sport Shoe	Tokyo	1	$1,155	$15,602	$22

Notice that the exterior line around the data is thicker than the interior lines of the data table. This indicates that GRIDLINES are turned 'ON' and that a blank line appears at the bottom. This blank line at the bottom does not get output to all Excel output.

The PAGES_FITHEIGHT= Suboption

This suboption is similar to the FITTOPAGE suboption in that it also adjusts the 'Fit to' information on the Excel 'Page Setup' sheet. This option adjusts the number of pages in the height box of the 'Fit to' Excel value. However, Excel is still in control of the printing, so you might have to experiment with this suboption to get Excel to output what you really want. Excel has a tendency to reduce the font size with this suboption.

SAS Code 7-15 – Using the PAGES_FITHEIGHT= Suboption to Put More Data on One Page

```
ods excel file = "&path\PAGES_FITHEIGHT_on.xlsx"
         options(PAGES_FITHEIGHT='2'
                 SHEET_INTERVAL='none');
   proc print data=Asia_only;
   run;
   proc print data=Asia_only;
   run;
ods excel close;
```

Figure 7-17 – The PAGES_FITHEIGHT Suboption Adjusts Excel FIT TO Number of Pages Tall

The PAGES_FITWIDTH= Suboption

This suboption is similar to the FITTOPAGE= suboption in that it also adjusts the 'Fit to' information on the Excel 'PAGE SETUP' sheet. PAGES_FITWIDTH= adjusts the number of pages in the width box of the 'Fit to' Excel value. However, Excel is still in control of the printing so you might have to experiment with this suboption to get Excel to output what you really want. Excel has a tendency to reduce the font size with this suboption.

SAS Code 7-16 – The Excel FIT TO 'Width' Option Puts More Information on One Printed Sheet

```
ods excel file = "&path\PAGES_FITWIDTH_on.xlsx"
           options(PAGES_FITWIDTH='2'
                   SHEET_INTERVAL='none');

   * print some variables three times to get a wide output;
   %let vars = region product subsidiary stores sales;
   proc print data=Asia_only;
      var &vars &vars &vars;
   run;

ods excel close;
```

Figure 7-18 – PAGES_FITWIDTH= Suboption Adjusts the Excel FIT TO Number of Pages Wide

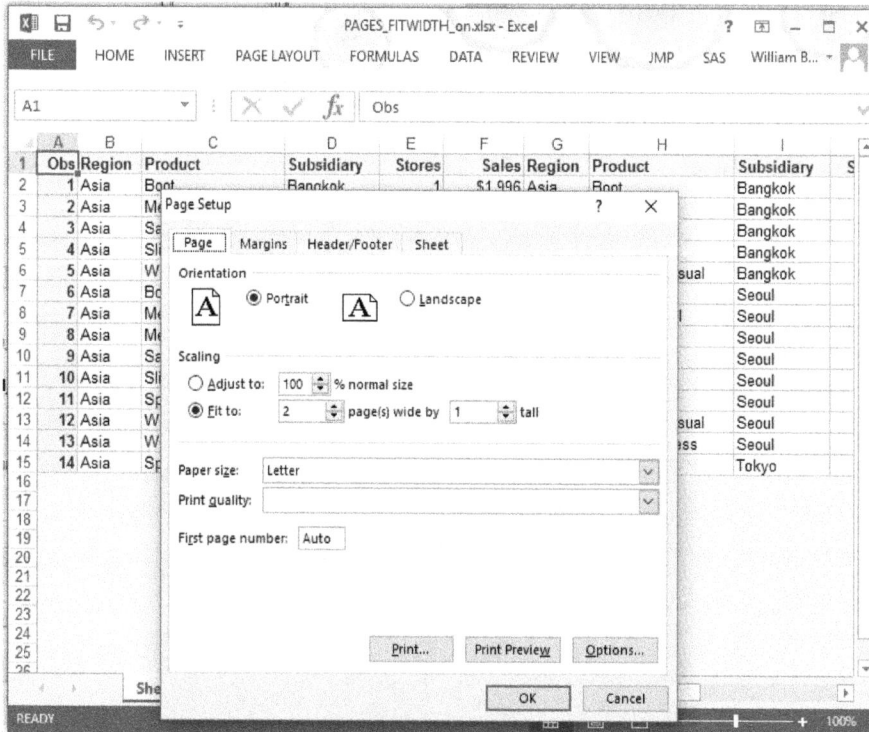

The SHEET_INTERVAL= Suboption

The SHEET_INTERVAL= suboption is useful when you have multiple outputs that you want to place into the same Excel Workbook. This suboption enables you to combine or separate the outputs in several different ways. There are five values that you can use with this suboption:

'BYGROUP', 'PAGE', 'PROC', 'NONE', and 'TABLE'. An example output of each of these values is shown separately below.

The suboption values perform the following tasks, with 'TABLE' being the default.

- 'BYGROUP' = a new worksheet is created for each new BYGROUP.
- 'PAGE' = a new worksheet is created for each new page of procedure output.
- 'PROC' = a new worksheet is created for each new procedure and it contains all data for that procedure.
- 'NONE' = all of the output is on one page.
- 'TABLE' = a new worksheet is created for each table.

SHEET_INTERVAL='BYGROUP' Example

When using the 'BYGROUP' value for the SHEET_INTERVAL= suboption, a new worksheet is created when the SAS 'BY GROUP' changes. The tab names of new BY GROUPS are generated by SAS, but other suboptions that deal with the sheet names can be used to customize the output sheet names. See the 'SHEET_LABEL=' and 'SHEET_NAME=' suboptions.

SAS Code 7-17 – Ways to Use the SHEET_INTERVAL= 'BYGROUP' Value

```
ods excel  file = "&path\SHEET_INTERVAL_bygroup.xlsx"
           options(SHEET_INTERVAL='bygroup');
   proc print data=sashelp.shoes;
      by region;
   run;
ods excel  close;
```

Using the 'BYGROUP' interval places each SAS BY Group onto a new Excel worksheet. See Figure 7-19 for an example. Default tab names were generated containing "By Group," the group number starting from one, and the By Group region value.

Figure 7-19 – SHEET_INTERVAL= 'BYGROUP' First Sheet of Excel Output

Figure 7-19 is the first sheet of output generated using the SHEET_INTERVAL= 'BYGROUP' value. Default tab names were generated and each worksheet contains data for a different Region value of the SASHELP.SHOES data set.

Figure 7-20 – SHEET_INTERVAL= 'BYGROUP' Second Sheet of Excel Output

Figure 7-20 is the second sheet of output generated using the SHEET_INTERVAL= 'BYGROUP' value. Default tab names were generated and each worksheet contains data for a different Region value of the SASHELP.SHOES data set.

SHEET_INTERVAL='PAGE' Example

When using the 'PAGE' value for the SHEET_INTERVAL= suboption, a new worksheet is created when a new page of output is generated. This is sometimes hard to show, here I used two procedures PROC PRINT and PROC UNIVARIATE to generate a new 'PAGE' of output.

SAS Code 7-18 – Ways to Use the SHEET_INTERVAL= 'PAGE' Value

```
ods excel   file = "&path\SHEET_INTERVAL_page.xlsx"
             options(SHEET_INTERVAL='page');
   proc print data=Asia_only;
   run;

   PROC UNIVARIATE   DATA=sashelp.shoes
      (keep =   region stores product
         where=(region="Asia" and product="Boot"));
   BY region;
   variable stores;
   freq stores;
   RUN;
ods excel   close;
```

In the output Excel workbook the tab names show the values 'PAGE' followed by a number and then the procedure name. The tab names are generated by SAS, but other suboptions that deal with the sheet names can be used to customize the output sheet names. See the 'SHEET_LABEL=' and 'SHEET_NAME=' suboptions.

Figure 7-21 – The ODS Excel SHEET_INTERVAL= Option 'PAGE' Page One Output

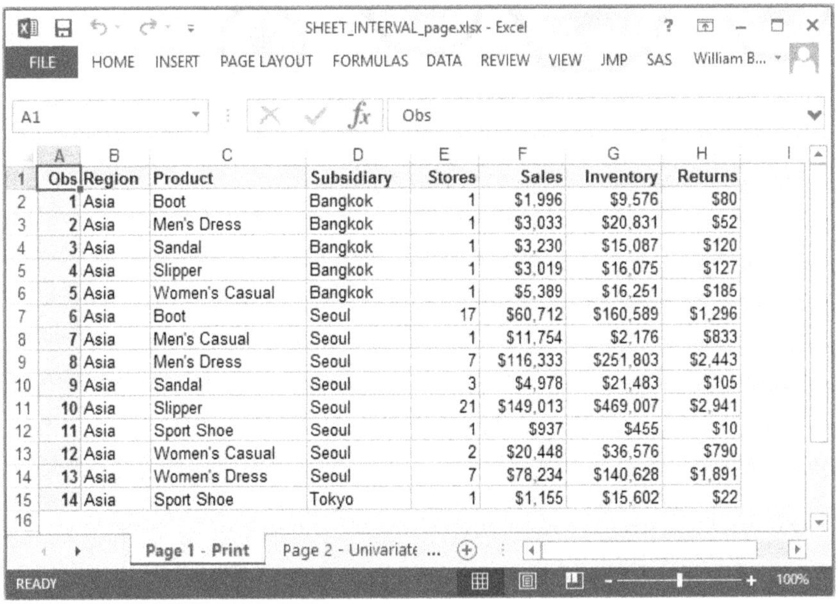

Page one is the PROC PRINT output of 14 lines of data.

Figure 7-22 – The ODS Excel SHEET_INTERVAL= Option 'PAGE' Page Two Output

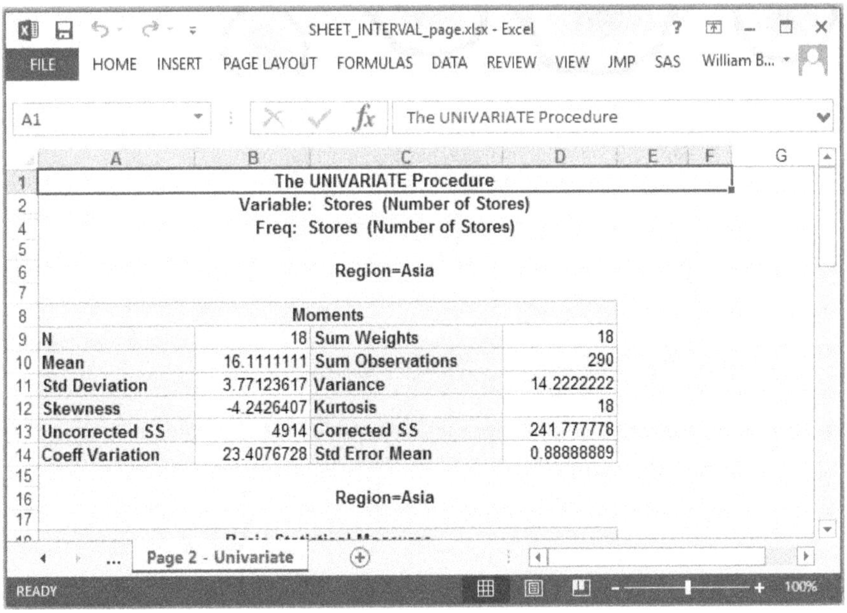

Page two is the full PROC UNIVARIATE output.

SHEET_INTERVAL='PROC' Example

When using the 'PROC' value for the SHEET_INTERVAL= suboption, a new worksheet is created when a new procedure executes and outputs data. This is easy to show, here I used two procedures PROC PRINT and PROC UNIVARIATE to generate a new worksheet when each procedure executes. In many cases this output looks the same as when the 'PAGE' value is used.

SAS Code 7-19 – Ways to Use the SHEET_INTERVAL= 'PROC' Value

```
ods excel   file = "&path\SHEET_INTERVAL_proc.xlsx"
           options(SHEET_INTERVAL='proc');
   proc print data=Asia_only;
   run;

   PROC UNIVARIATE  DATA=sashelp.shoes
      (keep =  region stores product
       where=(region="Asia" and product="Boot"));
   BY region;
   variable stores;
   freq stores;
   RUN;
ods excel  close;
```

Figure 7-23 – The SHEET_INTERVAL= Suboption Using the Value 'PROC', Page One

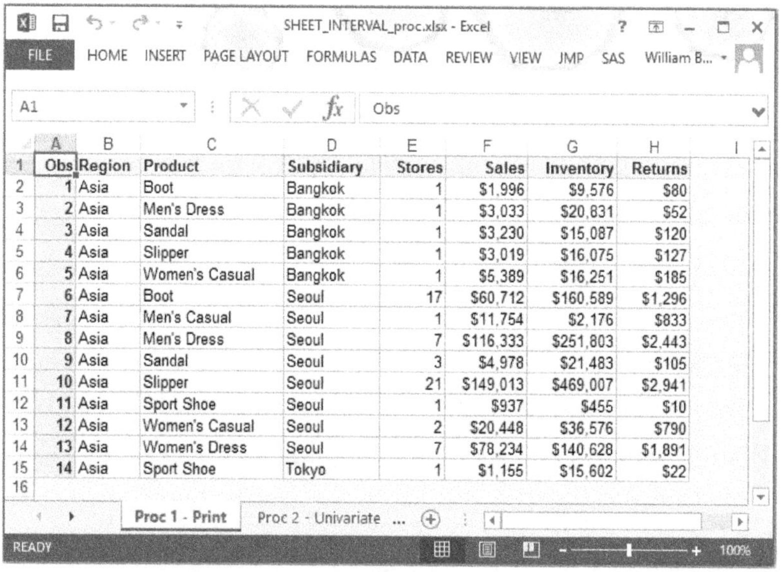

Obs	Region	Product	Subsidiary	Stores	Sales	Inventory	Returns
1	Asia	Boot	Bangkok	1	$1,996	$9,576	$80
2	Asia	Men's Dress	Bangkok	1	$3,033	$20,831	$52
3	Asia	Sandal	Bangkok	1	$3,230	$15,087	$120
4	Asia	Slipper	Bangkok	1	$3,019	$16,075	$127
5	Asia	Women's Casual	Bangkok	1	$5,389	$16,251	$185
6	Asia	Boot	Seoul	17	$60,712	$160,589	$1,296
7	Asia	Men's Casual	Seoul	1	$11,754	$2,176	$833
8	Asia	Men's Dress	Seoul	7	$116,333	$251,803	$2,443
9	Asia	Sandal	Seoul	3	$4,978	$21,483	$105
10	Asia	Slipper	Seoul	21	$149,013	$469,007	$2,941
11	Asia	Sport Shoe	Seoul	1	$937	$455	$10
12	Asia	Women's Casual	Seoul	2	$20,448	$36,576	$790
13	Asia	Women's Dress	Seoul	7	$78,234	$140,628	$1,891
14	Asia	Sport Shoe	Tokyo	1	$1,155	$15,602	$22

Notice here that the Excel worksheet names begin with the prefix 'Proc', which is the page type indicator, followed by a number and the indicator of the procedure name.

Figure 7-24 – The SHEET_INTERVAL= Suboption Using the Value 'PROC', Page Two

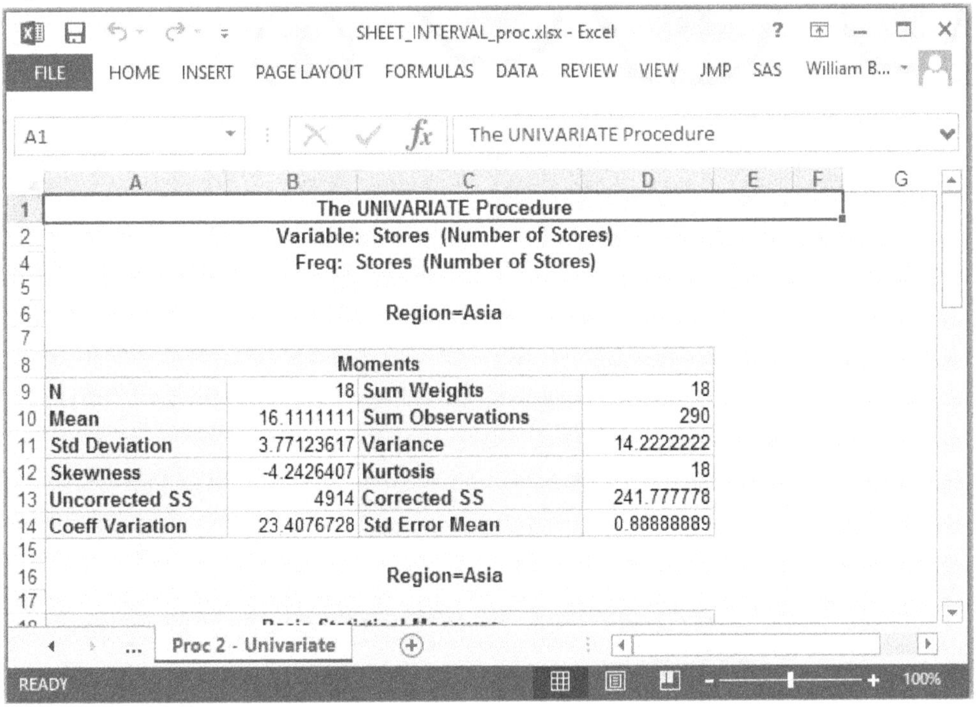

All of the data for both PROC PRINT and PROC UNIVARIATE appears in a separate worksheet for each procedure. The tab names are generated by SAS, but other suboptions that deal with the sheet names can be used to customize the output sheet names. See the 'SHEET_LABEL=' and 'SHEET_NAME=' suboptions.

SHEET_INTERVAL='NONE' Example

SAS Code 7-20 shows code using the 'NONE' value for the SHEET_INTERVAL= suboption where all data is sent to the same worksheet. No new worksheets are created until the SHEET_INTERVAL= value is changed for this workbook. This is easy to show, here I used two PROC PRINT procedures to generate a long listing. Figure 7-25 shows only one worksheet named 'Sheet 1'.

SAS Code 7-20 – Ways to Use the SHEET_INTERVAL= 'NONE' Value

```
ods excel  file = "&path\SHEET_INTERVAL_none.xlsx"
           options(SHEET_INTERVAL='none');
   proc print data=Asia_only;
   run;
   proc print data=Asia_only;
   run;
ods excel  close;
```

Figure 7-25 – The SHEET_INTERVAL Value 'NONE'

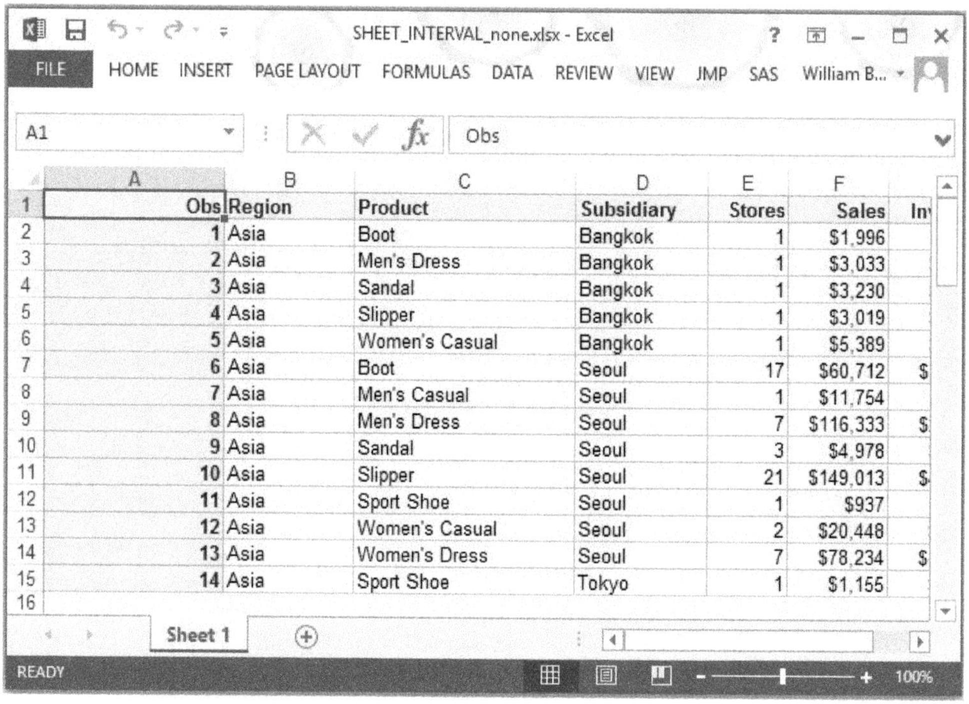

All output appears on one page when value 'NONE' is active.

SHEET_INTERVAL='TABLE' Example

SAS Code 7-21 shows code using the 'TABLE' value for the SHEET_INTERVAL= suboption. Data for each table is sent to a different worksheet, here I used two procedures PROC PRINT and PROC UNIVARIATE to generate many tables. See Figures 7-26 to 7-28.

SAS Code 7-21 – Ways to Use the SHEET_INTERVAL='TABLE' Value

```
ods excel  file = "&path\SHEET_INTERVAL_table.xlsx"
           options(SHEET_INTERVAL='table');
   proc print data=Asia_only;
   run;
   PROC UNIVARIATE  DATA=sashelp.shoes
      (keep =  region stores product
       where=(region="Asia" and product="Boot"));
   BY region;
   variable stores;
   freq stores;
   RUN;
ods excel  close;
```

Figure 7-26 – The First Table SHEET_INTERVAL

Obs	Region	Product	Subsidiary	Stores	Sales	Inventory	Returns
1	Asia	Boot	Bangkok	1	$1,996	$9,576	$80
2	Asia	Men's Dress	Bangkok	1	$3,033	$20,831	$52
3	Asia	Sandal	Bangkok	1	$3,230	$15,087	$120
4	Asia	Slipper	Bangkok	1	$3,019	$16,075	$127
5	Asia	Women's Casual	Bangkok	1	$5,389	$16,251	$185
6	Asia	Boot	Seoul	17	$60,712	$160,589	$1,296
7	Asia	Men's Casual	Seoul	1	$11,754	$2,176	$833
8	Asia	Men's Dress	Seoul	7	$116,333	$251,803	$2,443
9	Asia	Sandal	Seoul	3	$4,978	$21,483	$105
10	Asia	Slipper	Seoul	21	$149,013	$469,007	$2,941
11	Asia	Sport Shoe	Seoul	1	$937	$455	$10
12	Asia	Women's Casual	Seoul	2	$20,448	$36,576	$790
13	Asia	Women's Dress	Seoul	7	$78,234	$140,628	$1,891
14	Asia	Sport Shoe	Tokyo	1	$1,155	$15,602	$22

Print 1 - Data Set WORK.ASIA

Figure 7-27 – The Second Table SHEET_INTERVAL

The UNIVARIATE Procedure

Variable: Stores (Number of Stores)
Freq: Stores (Number of Stores)

Region=Asia

Moments

N	18	Sum Weights	18
Mean	16.1111111	Sum Observations	290
Std Deviation	3.77123617	Variance	14.2222222
Skewness	-4.2426407	Kurtosis	18
Uncorrected SS	4914	Corrected SS	241.777778
Coeff Variation	23.4076728	Std Error Mean	0.88888889

Univariate 2 - Moments   Univaria ...

Figures 7-26 and 7-27 show the first two tables that are generated. The tab names of the worksheets indicate the data source, but do not indicate the number of sheets. To show the number of worksheets there are several ways that it can be done. You can scroll through all of the sheets to find all six sheets, you could add a 'Table of Contents' or an 'Index' to the workbook, or if you are careful you can enter ALT/F11 at the same time to show the Visual Basic for Applications project desktop. Figure 7-28 shows, in the upper left corner, the Excel filename and the names of all of the worksheets in the workbook. But, use caution. If you have never seen this part of Excel before, it is easy to type something that will make your workbook unusable. Just take my word that this exists.

Figure 7-28 – Excel Visual Basic for Applications Desktop

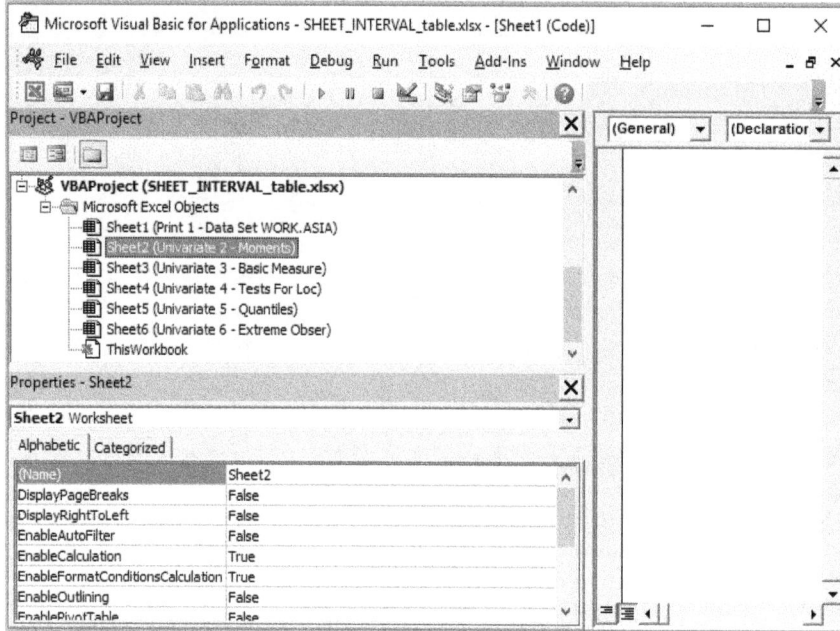

Be careful using the Visual Basic for Applications desktop, you can ruin your workbook here.

The SHEET_LABEL= Suboption

Since the default values for label names on the output sheets of an Excel workbook are not always the ones your boss wants to see, you might want to change the tab names. The first thing you can do is change the prefix of the generated tab names by using the ODS Excel SHEET_LABEL= option. When this option is not present or set to 'NONE', then SAS takes over the task of figuring out what names to apply to each tab of the workbook. You can use a 'string' of your choosing to change the prefix. This option only changes the prefix and allows SAS to determine the rest of the tab name. Excel tab names are 32 characters long, but SAS reserves the last four characters to ensure that unique names are output.

SAS Code 7-22 – ODS Excel SHEET_LABEL='NONE' Option Code

```
ods excel file = "&path\SHEET_LABEL_none.xlsx"
          options(SHEET_LABEL='none');
   proc print data=sashelp.shoes;
      by region;
   run;
ods excel close;
```

Figure 7-29 – Default Value of the SHEET_LABEL= 'NONE'

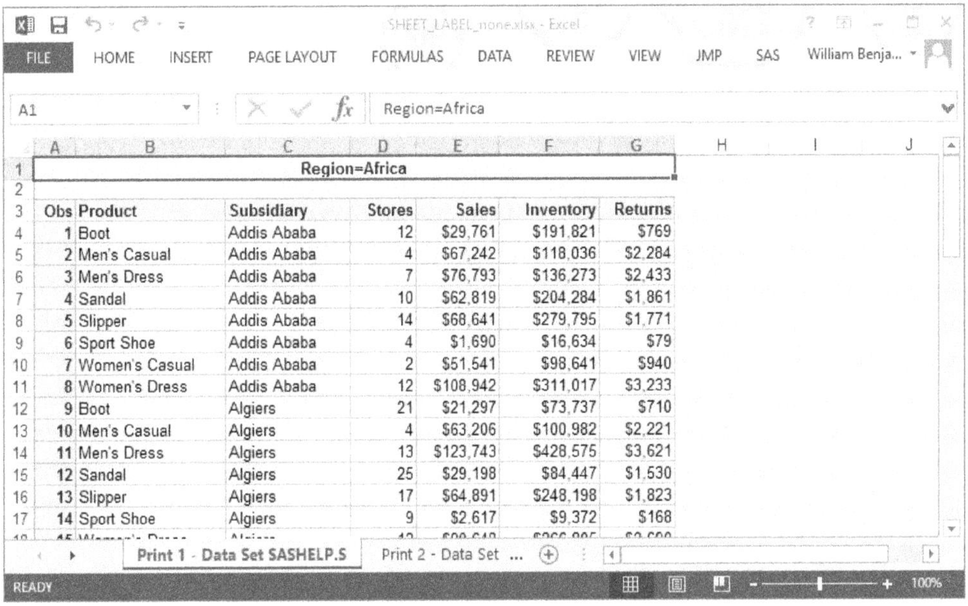

Region=Africa

Obs	Product	Subsidiary	Stores	Sales	Inventory	Returns
1	Boot	Addis Ababa	12	$29,761	$191,821	$769
2	Men's Casual	Addis Ababa	4	$67,242	$118,036	$2,284
3	Men's Dress	Addis Ababa	7	$76,793	$136,273	$2,433
4	Sandal	Addis Ababa	10	$62,819	$204,284	$1,861
5	Slipper	Addis Ababa	14	$68,641	$279,795	$1,771
6	Sport Shoe	Addis Ababa	4	$1,690	$16,634	$79
7	Women's Casual	Addis Ababa	2	$51,541	$98,641	$940
8	Women's Dress	Addis Ababa	12	$108,942	$311,017	$3,233
9	Boot	Algiers	21	$21,297	$73,737	$710
10	Men's Casual	Algiers	4	$63,206	$100,982	$2,221
11	Men's Dress	Algiers	13	$123,743	$428,575	$3,621
12	Sandal	Algiers	25	$29,198	$84,447	$1,530
13	Slipper	Algiers	17	$64,891	$248,198	$1,823
14	Sport Shoe	Algiers	9	$2,617	$9,372	$168

The default prefix for worksheets generated with PRINT procedure is "Print" followed by a number and some part of the SAS data set name. Other procedures will produce a different prefix. SAS Code 7-23 shows that you can use any string.

SAS Code 7-23 – ODS Excel SHEET_LABEL='My_Prefix' Value

```
ods excel file = "&path\SHEET_LABEL_string.xlsx"
          options(SHEET_LABEL='My_Prefix');
   proc print data=sashelp.shoes;
      by region;
   run;
ods excel close;
```

Figure 7-30 – Use of a 'string' with the SHEET_LABEL= Option to Change the Sheet Tab Prefix

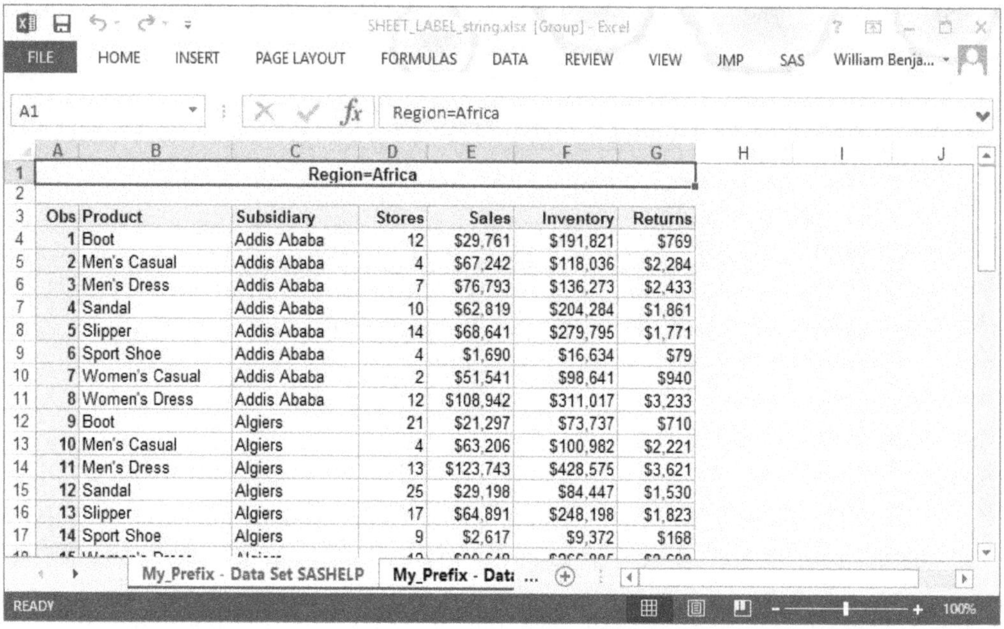

In Figure 7-30 the sheet tab prefix is "My_Prefix" but the rest of the tab name is the same as the default sheet label.

The SHEET_NAME= Suboption

This option, unlike the SHEET_LABEL= option, changes the whole sheet name. Of course, the default here is to not use the option, because any string used with this option replaces the whole sheet (tab) name. Once again, the Excel tab names can be 32 characters, but SAS reserves the last four characters for a unique number to ensure that the output sheets have unique names. The string that you provide is used for all of the tabs until the SHEET_NAME suboption is changed. The example shown in Figure 7-31 has multiple worksheets generated from the SAS BY statement, and the sheet names have a number following the name in the second and following worksheets. This suboption also accepts the SAS constant '#BYVAL' when you want to use the BY GROUP values as the tab names. Using this suboption enables you to have more control over the tab names of your Excel workbook.

SAS Code 7-24 – Using the SHEET_NAME= Option to Change Full Names of the Excel Worksheets

```
ods excel file = "&path\SHEET_NAME_My_String.xlsx"
          options(SHEET_NAME='My_Sheet_Names');
   proc print data=sashelp.shoes;
      by region;
   run;
ods excel close;
```

Figure 7-31 – Excel Output with Sheet Names Changed by the ODS Excel SHEET_NAME Option

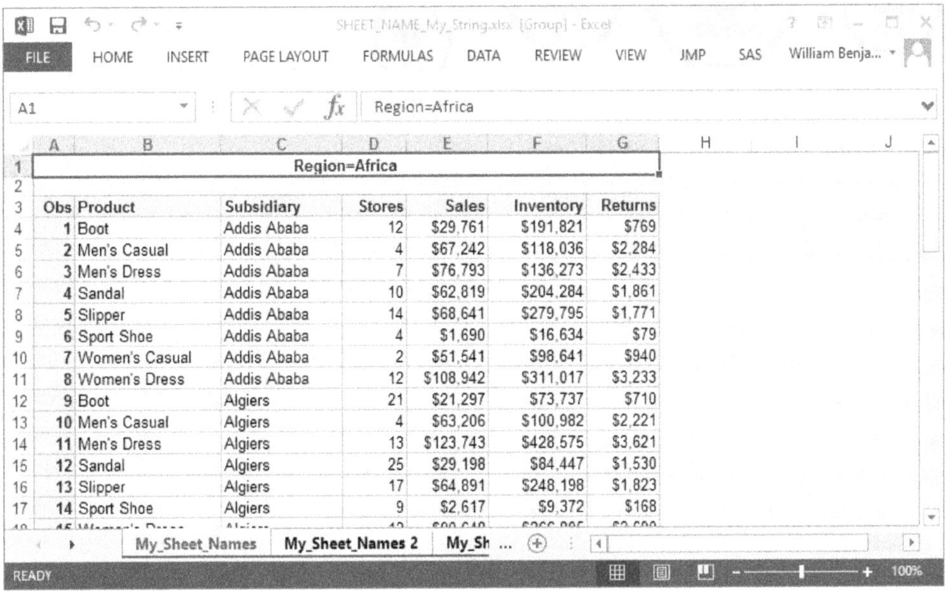

Region=Africa

Obs	Product	Subsidiary	Stores	Sales	Inventory	Returns
1	Boot	Addis Ababa	12	$29,761	$191,821	$769
2	Men's Casual	Addis Ababa	4	$67,242	$118,036	$2,284
3	Men's Dress	Addis Ababa	7	$76,793	$136,273	$2,433
4	Sandal	Addis Ababa	10	$62,819	$204,284	$1,861
5	Slipper	Addis Ababa	14	$68,641	$279,795	$1,771
6	Sport Shoe	Addis Ababa	4	$1,690	$16,634	$79
7	Women's Casual	Addis Ababa	2	$51,541	$98,641	$940
8	Women's Dress	Addis Ababa	12	$108,942	$311,017	$3,233
9	Boot	Algiers	21	$21,297	$73,737	$710
10	Men's Casual	Algiers	4	$63,206	$100,982	$2,221
11	Men's Dress	Algiers	13	$123,743	$428,575	$3,621
12	Sandal	Algiers	25	$29,198	$84,447	$1,530
13	Slipper	Algiers	17	$64,891	$248,198	$1,823
14	Sport Shoe	Algiers	9	$2,617	$9,372	$168

The SUPPRESS_BYLINES= Suboption

The default value for the SUPPRESS_BYLINES= suboption is to not suppress them. Figure 7-32 shows this condition. The SHEET_INTERVAL= 'NONE' suboption forced the output from the PRINT procedure to one worksheet so that the display would fit into one image.

SAS Code 7-25 – Using the SUPPRESS_BYLINES='OFF' Value

```
ods excel file = "&path\SUPPRESS_BYLINES_off.xlsx"
          options(SUPPRESS_BYLINES='off'
                  SHEET_INTERVAL='none');
   proc print data=sashelp.shoes
                  (where=(region ne 'Africa'));
      by region;
   run;
ods excel close;
```

This code not only forces all of the output data to one Excel worksheet, it also does not include the data for region 'Africa' because that data would not fit onto the first page, as shown in Figure 7-32.

Figure 7-32 – Excel Output Showing the 'BYLINE' Line Values

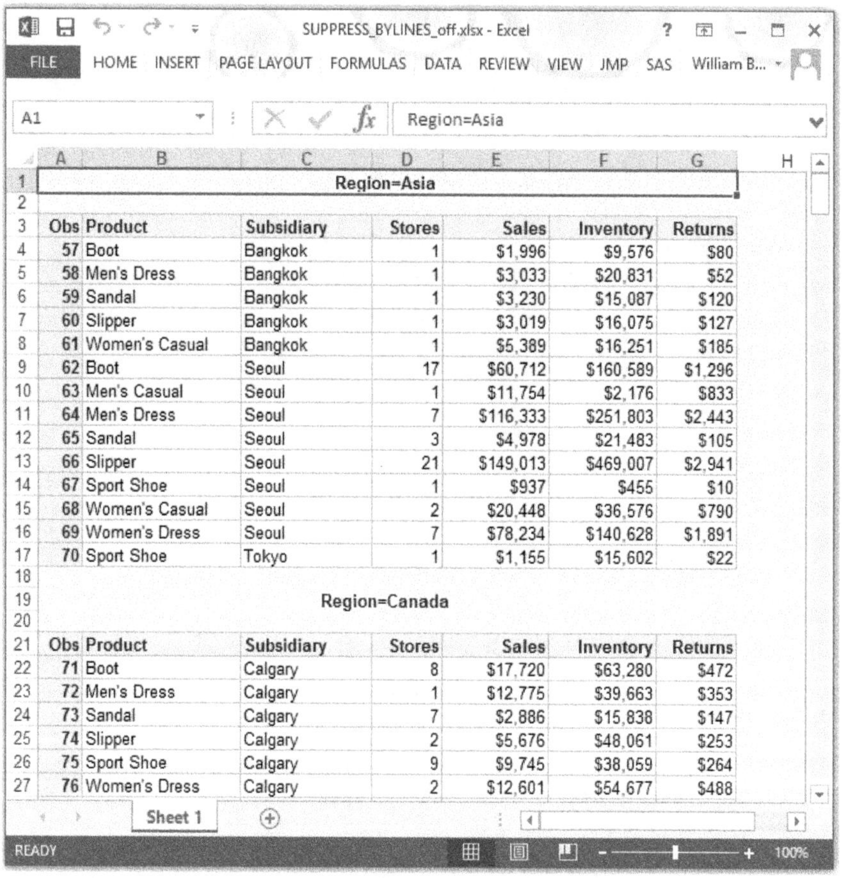

Region=Asia

Obs	Product	Subsidiary	Stores	Sales	Inventory	Returns
57	Boot	Bangkok	1	$1,996	$9,576	$80
58	Men's Dress	Bangkok	1	$3,033	$20,831	$52
59	Sandal	Bangkok	1	$3,230	$15,087	$120
60	Slipper	Bangkok	1	$3,019	$16,075	$127
61	Women's Casual	Bangkok	1	$5,389	$16,251	$185
62	Boot	Seoul	17	$60,712	$160,589	$1,296
63	Men's Casual	Seoul	1	$11,754	$2,176	$833
64	Men's Dress	Seoul	7	$116,333	$251,803	$2,443
65	Sandal	Seoul	3	$4,978	$21,483	$105
66	Slipper	Seoul	21	$149,013	$469,007	$2,941
67	Sport Shoe	Seoul	1	$937	$455	$10
68	Women's Casual	Seoul	2	$20,448	$36,576	$790
69	Women's Dress	Seoul	7	$78,234	$140,628	$1,891
70	Sport Shoe	Tokyo	1	$1,155	$15,602	$22

Region=Canada

Obs	Product	Subsidiary	Stores	Sales	Inventory	Returns
71	Boot	Calgary	8	$17,720	$63,280	$472
72	Men's Dress	Calgary	1	$12,775	$39,663	$353
73	Sandal	Calgary	7	$2,886	$15,838	$147
74	Slipper	Calgary	2	$5,676	$48,061	$253
75	Sport Shoe	Calgary	9	$9,745	$38,059	$264
76	Women's Dress	Calgary	2	$12,601	$54,677	$488

The data from 'Africa' was suppressed by the SAS code, and the 'BYLINE' values show up as 'Region =Asia', 'Region=Canada', and in all of the other regions in the SASHELP.SHOES data set.

SAS Code 7-26 – Using the SUPPRESS_BYLINES='ON' Value

```
ods excel file = "&path\SUPPRESS_BYLINES_on.xlsx"
          options(SUPPRESS_BYLINES='on'
                  SHEET_INTERVAL='none');
   proc print data=sashelp.shoes
                  (where=(region ne 'Africa'));
      by region;
   run;
ods excel close;
```

Figure 7-33 – Excel Output with BYLINES Suppressed

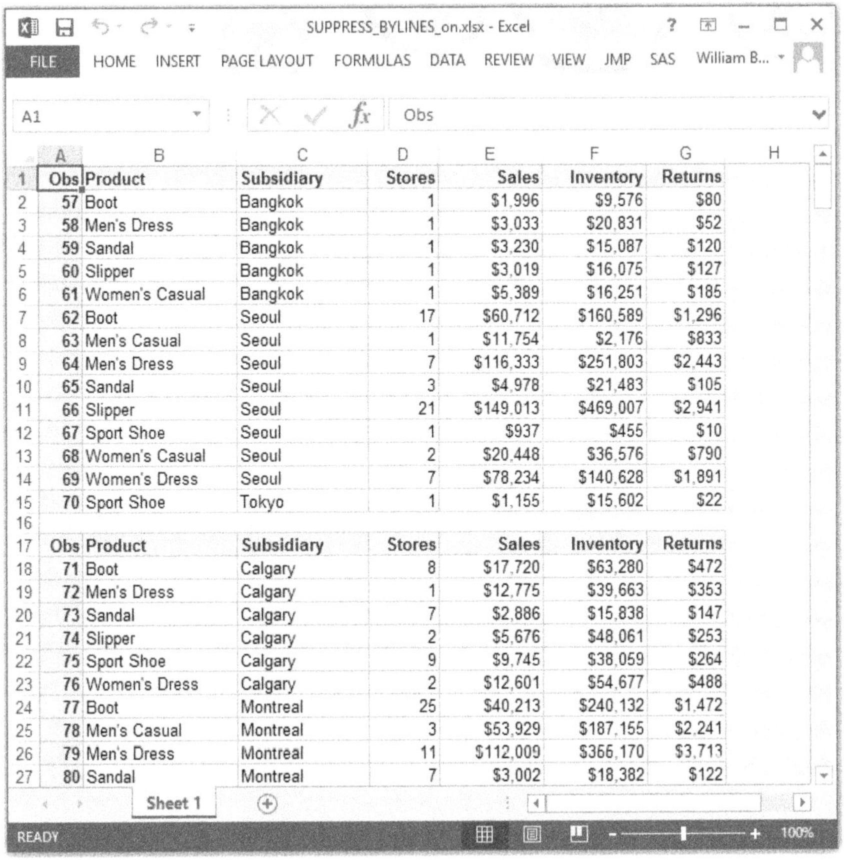

	A	B	C	D	E	F	G
1	Obs	Product	Subsidiary	Stores	Sales	Inventory	Returns
2	57	Boot	Bangkok	1	$1,996	$9,576	$80
3	58	Men's Dress	Bangkok	1	$3,033	$20,831	$52
4	59	Sandal	Bangkok	1	$3,230	$15,087	$120
5	60	Slipper	Bangkok	1	$3,019	$16,075	$127
6	61	Women's Casual	Bangkok	1	$5,389	$16,251	$185
7	62	Boot	Seoul	17	$60,712	$160,589	$1,296
8	63	Men's Casual	Seoul	1	$11,754	$2,176	$833
9	64	Men's Dress	Seoul	7	$116,333	$251,803	$2,443
10	65	Sandal	Seoul	3	$4,978	$21,483	$105
11	66	Slipper	Seoul	21	$149,013	$469,007	$2,941
12	67	Sport Shoe	Seoul	1	$937	$455	$10
13	68	Women's Casual	Seoul	2	$20,448	$36,576	$790
14	69	Women's Dress	Seoul	7	$78,234	$140,628	$1,891
15	70	Sport Shoe	Tokyo	1	$1,155	$15,602	$22
16							
17	Obs	Product	Subsidiary	Stores	Sales	Inventory	Returns
18	71	Boot	Calgary	8	$17,720	$63,280	$472
19	72	Men's Dress	Calgary	1	$12,775	$39,663	$353
20	73	Sandal	Calgary	7	$2,886	$15,838	$147
21	74	Slipper	Calgary	2	$5,676	$48,061	$253
22	75	Sport Shoe	Calgary	9	$9,745	$38,059	$264
23	76	Women's Dress	Calgary	2	$12,601	$54,677	$488
24	77	Boot	Montreal	25	$40,213	$240,132	$1,472
25	78	Men's Casual	Montreal	3	$53,929	$187,155	$2,241
26	79	Men's Dress	Montreal	11	$112,009	$366,170	$3,713
27	80	Sandal	Montreal	7	$3,002	$18,382	$122

The data from 'Africa' was suppressed by the SAS code, and the 'BYLINE' values are also suppressed in this Excel workbook.

Conclusion

The examples in this chapter help to clearly visualize how you can adjust the ways that titles, footnotes, and label information is displayed on your output worksheet. Having the ability to introduce these features into your Excel output at the time that your SAS code runs can be a huge time saving coding practice. Proper use of these features can also enhance the way that your final workbooks appear, without having to modify your workbooks after they are generated.

Chapter 8: Options That Affect Print Features

Introduction

This chapter will discuss the ODS Excel destination suboptions that affect output at the level of the whole workbook. Some of these suboptions affect the color, size, position, spacing, and visibility of the output data in the Excel workbook. The manipulation of headers, titles, footers, footnotes, spacing, and print order are often things that the creator of the workbook must do after the completion of the SAS program to build the workbook. These features of the ODS Excel destination allow these tasks to be accomplished as the Excel workbook is being created. By being able to create the workbook with these features installed you can deliver a workbook that has not been modified.

ODS Excel Destination Actions

In this chapter we will discuss the following SAS ODS EXCEL destination topics.

Table 8-1 – ODS Excel Destination Actions

Action Parameter	Options	Description
BLACKANDWHITE	'OFF', 'ON', 'YES', 'NO' Default='OFF'	Positive responses change the print mode to black and white printing only, negative responses allow color printing.
CENTER_HORIZONTAL	'OFF', 'ON', 'YES', 'NO' Default='OFF'	Positive responses center the worksheet horizontally on the printed page, negative responses do not horizontally center the printed output.
CENTER_VERTICAL	'OFF', 'ON', 'YES', 'NO' Default='OFF'	Positive responses center the worksheet vertically on the printed page, negative responses do not vertically center the printed output.
DRAFTQUALITY	'OFF', 'ON', 'YES', 'NO' Default='OFF'	Positive responses for this option will cause the output to be printed in "DRAFT" quality mode. However, graphs will not be printed when this option is in effect.
ORIENTATION	'PORTRAIT', 'LANDSCAPE', Default= 'PORTRAIT'	This option aligns the printed output as either PORTRAIT or LANDSCAPE on the printed page.
PAGE_ORDER_ACROSS	'OFF', 'ON', 'YES', 'NO' Default='OFF'	This option controls what order the information in the spreadsheet is printed. Positive responses cause the data to be printed across the page first, and then down the page. Negative responses cause the printing to be down the page first then across the page.
PRINT_AREA	'ITEM', Default= NONE	This option enables you to define an output print area that is a subset of the values (cells) of the spreadsheet. 'ITEM' is a quoted string of values that represent the top left cell and the bottom right cell of the spreadsheet to be printed. An example would be 'b,4,f,10'.
PRINT_FOOTER	'text-string'	This text string will be placed in the footer when printing, unless a footnote is active. This does not override the FOOTNOTE statement.
PRINT_FOOTER_MARGIN	'number',	This option specifies the amount of the

Action Parameter	Options	Description
	Default=0.5 inches	footer margin, and is set in the Excel page setup window. The units specified are in inches.
PRINT_HEADER	'text-string'	This text string will be placed in the header when printing, unless a title is active. This does not override the TITLE statement.
PRINT_HEADER_MARGIN	'number', Default= 0.5 inches	This option specifies the amount of the header margin, and is set in the Excel page setup window. The units specified are in inches.
ROWBREAKS_COUNT	'number'	This option specifies that a page break will be inserted after the number of data lines specified are printed.
ROWBREAKS_INTERVAL	'OUTPUT', 'PROC', 'NONE' Default='NONE'	This option controls the placement of page breaks. • 'OUTPUT' puts a page break between each output object. • 'PROC' puts a page break between each procedure output. • 'NONE' does not insert custom page breaks.
ROWCOLHEADINGS	'OFF', 'ON', 'YES', 'NO' Default='OFF'	Positive responses indicate that row and column headings should be printed. Negative responses indicate that row and column headings should not be printed.
SCALE	'number', Default= 100	This option controls the scaling of the output printout.
TITLE_FOOTNOTE_NOBREAK	'OFF', 'ON', 'YES', 'NO' Default='NO'	This option controls whether the titles and footnotes wrap across lines. Positive responses suppress wrapping, while negative responses allow wrapping. Think "Yes, we have no bananas today."
TITLE_FOOTNOTE_WIDTH	'number', Default='0'	This option specifies the number of columns that the titles and footnotes span. If the value is '0' (zero), then the titles and footnotes span across all columns that are in use in the spreadsheet.

The BLACKANDWHITE= Option

This ODS Excel option enables you to force the print output to be in "Black and White" only. This is useful if your printer does not print color images. Sometimes the color tones fade to all the same color if printed on a black and white printer. Hopefully this will help smooth them out. I did not show the default here, which is 'OFF'.

SAS Code 8-1 – The BLACKANDWHITE= Option Turned 'ON'

```
ods excel file = "&path\Black_and_white_printing.xlsx"
           options(BLACKANDWHITE='ON');
   proc print data=Asia_only;
   run;
ods excel close;
```

Figure 8-1 – The BLACKANDWHITE= Option Set to 'ON'

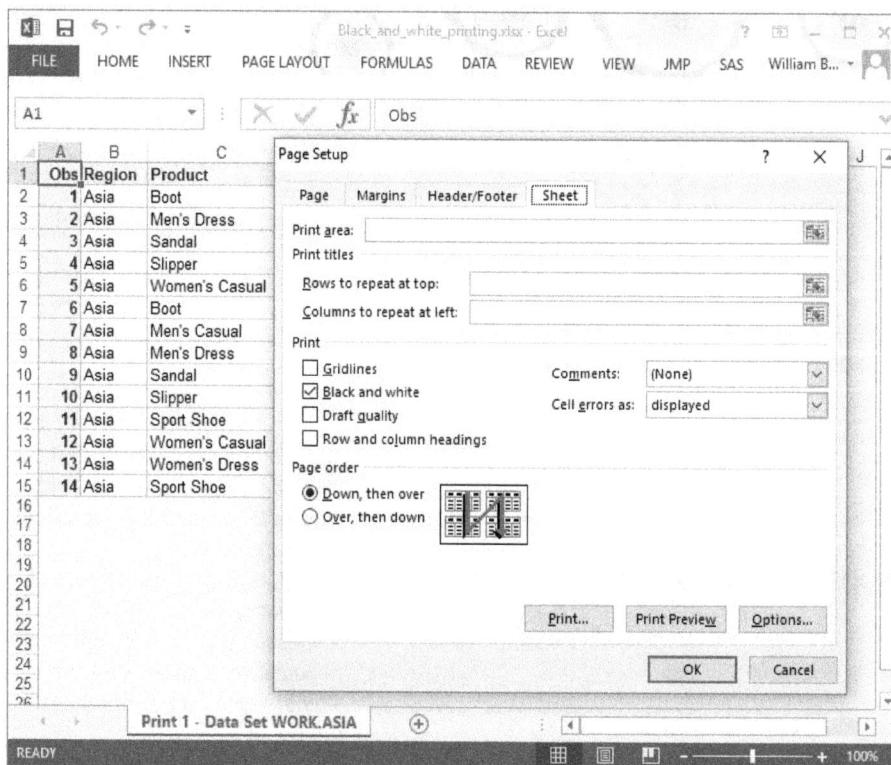

Under the "Print" section the check box for Black and White is selected.

The CENTER_HORIZONTAL= Option

This option sets the Excel 'Page Setup' indicator 'Center on Page – Horizontally' to active when the option = 'ON'. The result is that the output data is either centered between the left and right sides of the output page or is it not centered. 'ON' centers the output data. Figure 8-2 shows the option turned off. The print images in Figure 8-3 and Figure 8-5 were executed in 'Portrait' mode, but cut short to conserve space.

SAS Code 8-2 – The CENTER_HORIZONTAL= Options OFF and ON

```
ods excel (id=1) file = "&path\CENTER_HORIZONTAL_printing_off.xlsx"
          options(CENTER_HORIZONTAL='OFF');
ods excel (id=2) file = "&path\CENTER_HORIZONTAL_printing_on.xlsx"
          options(CENTER_HORIZONTAL='ON');
     proc print data=Asia_only;
   run;
ods excel (id=1) close;
ods excel (id=2) close;
```

Figure 8-2 – CENTER_HORIZONTALLY='OFF' Page Setup Sheet

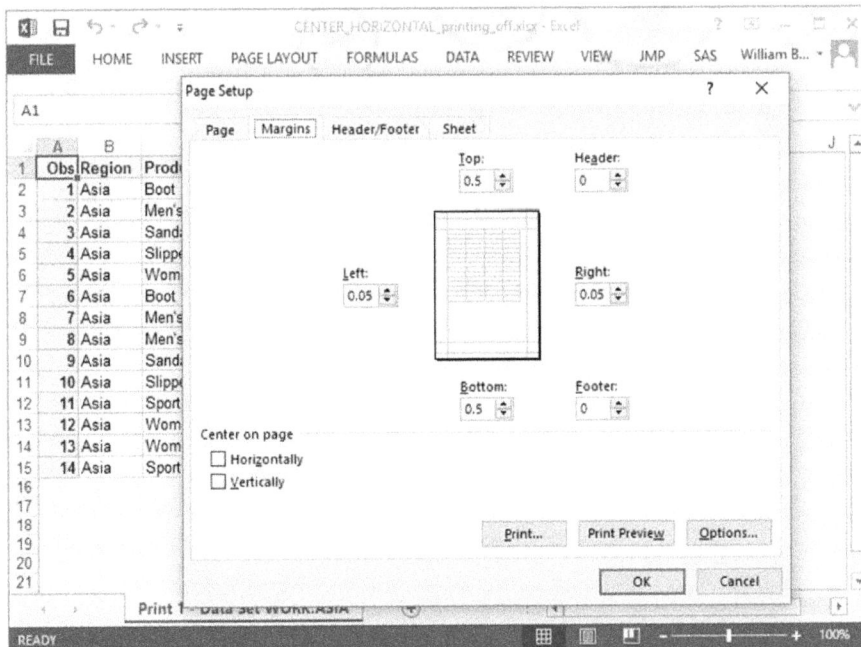

In the 'Center on page' area the check box for 'Horizontally' is not checked. And the margin image shows the output left-justified.

Figure 8-3 – Printed Output That is Not Centered

The SAS System

Obs	Region	Product	Subsidiary	Stores	Sales	Inventory	Returns
1	Asia	Boot	Bangkok	1	$1,996	$9,576	$80
2	Asia	Men's Dress	Bangkok	1	$3,033	$20,831	$52
3	Asia	Sandal	Bangkok	1	$3,230	$15,087	$120
4	Asia	Slipper	Bangkok	1	$3,019	$16,075	$127
5	Asia	Women's Casual	Bangkok	1	$5,389	$16,251	$185
6	Asia	Boot	Seoul	17	$80,712	$160,589	$1,296
7	Asia	Men's Casual	Seoul	1	$11,754	$2,176	$833
8	Asia	Men's Dress	Seoul	7	$116,333	$251,803	$2,443
9	Asia	Sandal	Seoul	3	$4,978	$21,483	$105
10	Asia	Slipper	Seoul	21	$149,013	$469,007	$2,941
11	Asia	Sport Shoe	Seoul	1	$937	$455	$10
12	Asia	Women's Casual	Seoul	2	$20,448	$36,576	$790
13	Asia	Women's Dress	Seoul	7	$78,234	$140,828	$1,891
14	Asia	Sport Shoe	Tokyo	1	$1,155	$15,602	$22

This is the default output style printed in portrait mode with the page shortened to conserve space.

Figure 8-4 – CENTER_HORIZONTAL=‘ON’ Excel Settings

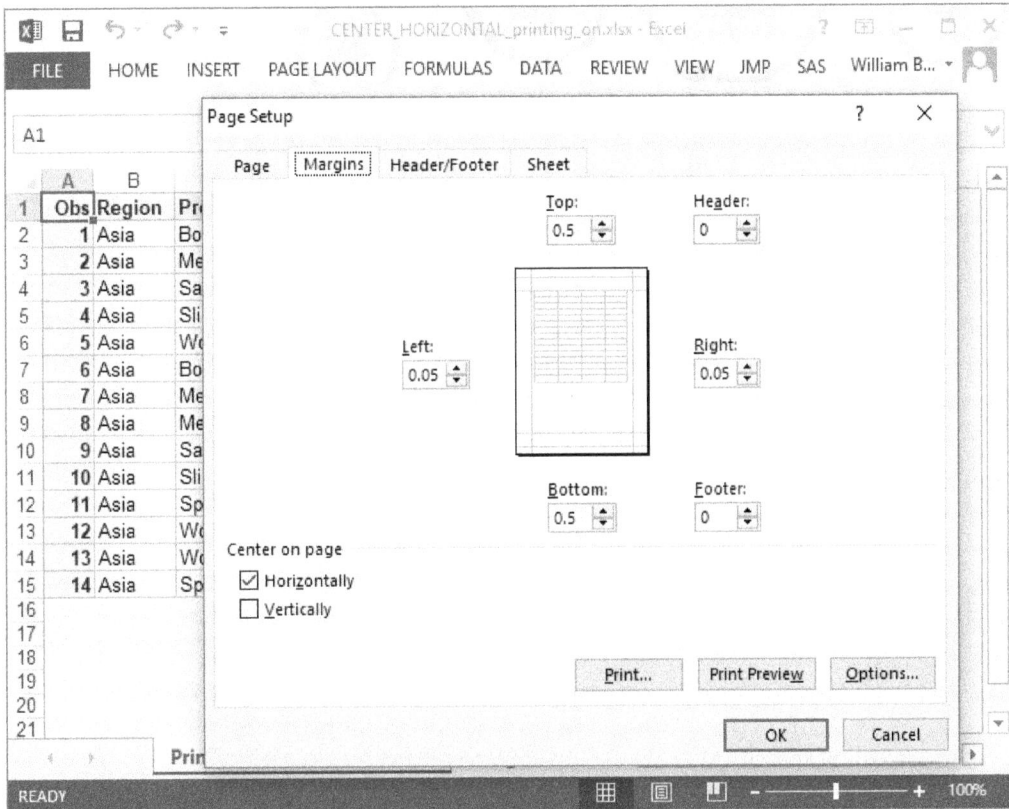

In the 'Center on page' area the check box for 'Horizontally' is checked. And the margin image shows the output centered.

Figure 8-5 – Centered Output from the CENTER_HORIZONTAL='ON' Option

The SAS System

Obs	Region	Product	Subsidiary	Stores	Sales	Inventory	Returns
1	Asia	Boot	Bangkok	1	$1,996	$9,576	$80
2	Asia	Men's Dress	Bangkok	1	$3,033	$20,831	$52
3	Asia	Sandal	Bangkok	1	$3,230	$15,087	$120
4	Asia	Slipper	Bangkok	1	$3,019	$16,075	$127
5	Asia	Women's Casual	Bangkok	1	$5,389	$16,251	$185
6	Asia	Boot	Seoul	17	$60,712	$160,589	$1,296
7	Asia	Men's Casual	Seoul	1	$11,754	$2,176	$833
8	Asia	Men's Dress	Seoul	7	$116,333	$251,803	$2,443
9	Asia	Sandal	Seoul	3	$4,978	$21,483	$105
10	Asia	Slipper	Seoul	21	$149,013	$469,007	$2,941
11	Asia	Sport Shoe	Seoul	1	$937	$455	$10
12	Asia	Women's Casual	Seoul	2	$20,448	$36,576	$790
13	Asia	Women's Dress	Seoul	7	$78,234	$140,628	$1,891
14	Asia	Sport Shoe	Tokyo	1	$1,155	$15,602	$22

This page was printed in portrait mode but the page was shortened to conserve space.

The CENTER_VERTICAL= Option

This option sets the Excel 'Page Setup' indicator 'Center on Page –Vertically' to active when the option = 'ON'. The result is that the output data is either centered between the top and bottom edges of the output page or is it not centered. 'ON' centers the output data. Figure 8-6 shows the option turned off. The print images for CENTER_VERTICAL='OFF' are the same as in Figure 8-2 and Figure 8-3 and are not repeated here.

SAS Code 8-3 – The CENTER_VERTICAL= Options OFF and ON

```
ods excel (id=1) file = "&path\CENTER_VERTICAL_printing_off.xlsx"
          options(CENTER_VERTICAL='OFF');
ods excel (id=2) file = "&path\CENTER_VERTICAL_printing_on.xlsx"
          options(CENTER_VERTICAL='ON');

   proc print data=Asia_only;
   run;
ods excel (id=1) close;
ods excel (id=2) close;
```

Figure 8-6 – CENTER_HORIZONTAL='ON' Excel Settings

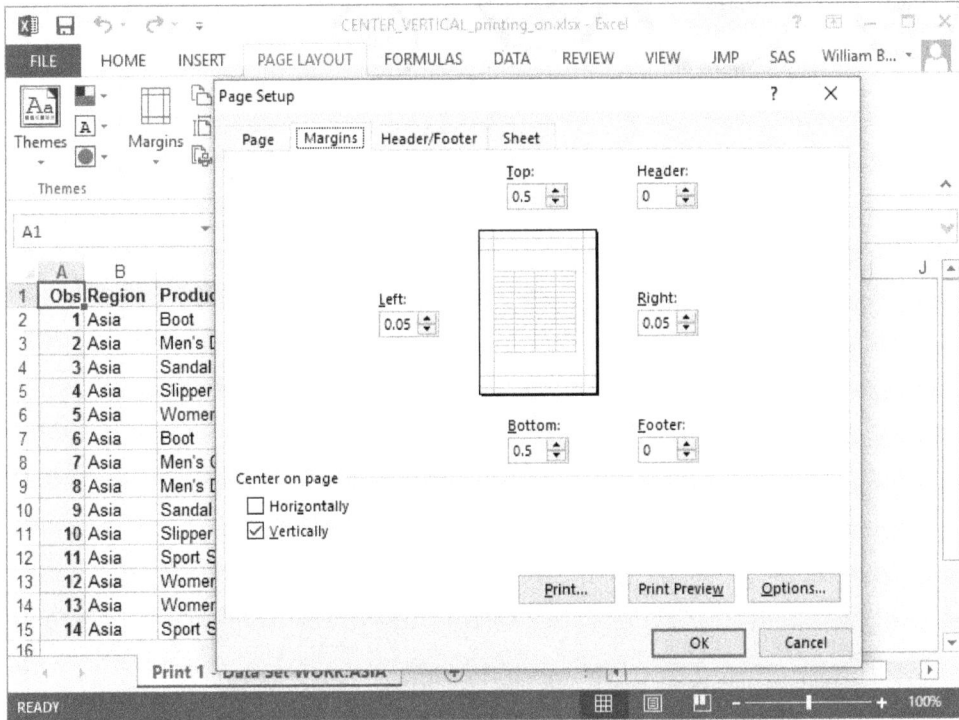

The 'Center on page – Vertically' check box is marked.

Figure 8-7 – Centered Output from the CENTER_HORIZONTAL='ON' Option

The SAS System

Obs	Region	Product	Subsidiary	Stores	Sales	Inventory	Returns
1	Asia	Boot	Bangkok	1	$1,996	$9,576	$80
2	Asia	Men's Dress	Bangkok	1	$3,033	$20,831	$52
3	Asia	Sandal	Bangkok	1	$3,230	$15,087	$120
4	Asia	Slipper	Bangkok	1	$3,019	$16,075	$127
5	Asia	Women's Casual	Bangkok	1	$5,389	$16,251	$185
6	Asia	Boot	Seoul	17	$60,712	$160,589	$1,296
7	Asia	Men's Casual	Seoul	1	$11,754	$2,176	$833
8	Asia	Men's Dress	Seoul	7	$116,333	$251,803	$2,443
9	Asia	Sandal	Seoul	3	$4,978	$21,483	$105
10	Asia	Slipper	Seoul	21	$149,013	$469,007	$2,941
11	Asia	Sport Shoe	Seoul	1	$937	$455	$10
12	Asia	Women's Casual	Seoul	2	$20,448	$36,576	$790
13	Asia	Women's Dress	Seoul	7	$78,234	$140,828	$1,891
14	Asia	Sport Shoe	Tokyo	1	$1,155	$15,602	$22

The data is centered on the page between the top and bottom edges of the page margins. Take note that the horizontal centering was turned off here, so the page is left-justified.

The DRAFTQUALITY= Option

I do a lot of work from home, and buy my own paper and ink. I use a good printer, but it is not really a fast printer. I use the draft mode when I print as often as I can to save ink. This setting allows me to preset the draft mode for printing so that I do not forget to turn it on when I actually do print an image. The default setting is 'OFF" and is not shown here. Since I cannot show a distinction here in the book, I am showing the setting.

SAS Code 8-4 – Setting the DRAFTQUALITY= Option to 'ON'

```
ods excel file = "&path\DRAFTQUALITY_printing_on.xlsx"
           options(DRAFTQUALITY='ON');
   proc print data=Asia_only;
   run;
ods excel close;
```

Figure 8-8 – Excel Page Setup Screen Showing the DRAFTQUALITY= Option Set to 'ON'

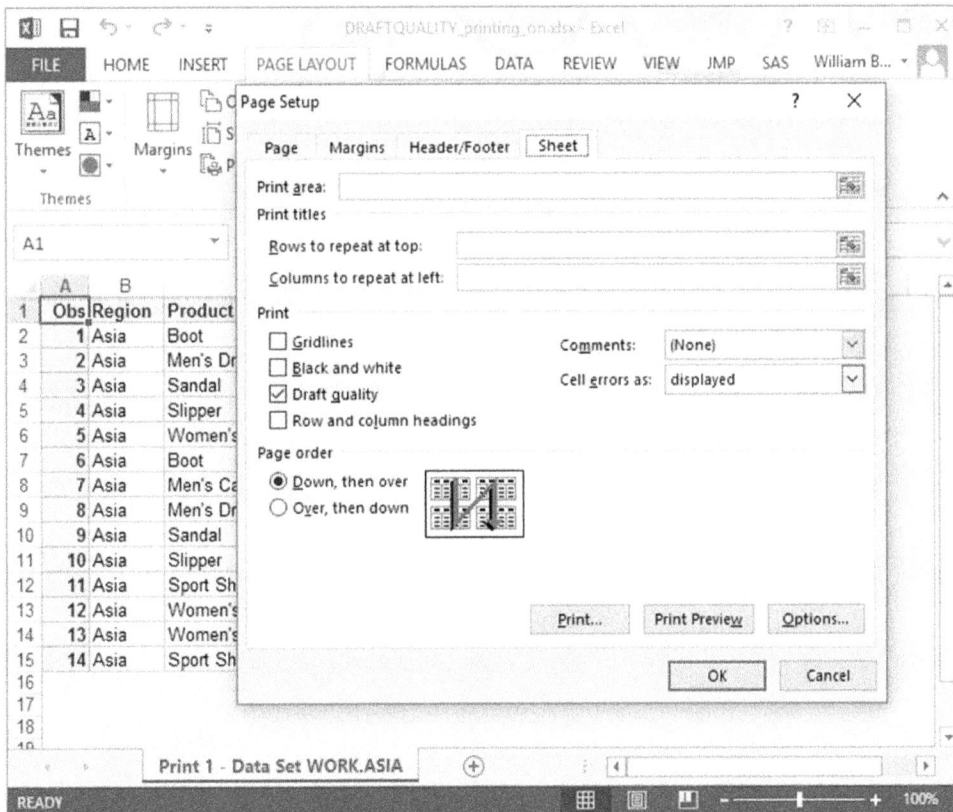

Under the 'Print' topic the 'Draft quality' check box is selected.

The ORIENTATION= Option

The ODS Excel ORIENTATION= option enables you to change how the printout appears on the page. The default is 'PORTRAIT' and is not shown here because most of the other images are also in the portrait mode. The options are 'PORTRAIT' or 'LANDSCAPE' and the Excel setting is shown on the 'Page Setup' sheet of the Excel worksheet.

SAS Code 8-5 – Setting the Printed Page Orientation to LANDSCAPE

```
ods excel file = "&path\ORIENTATION_LANDSCAPE_printing_on.xlsx"
          options(ORIENTATION='LANDSCAPE' );

   proc print data=Asia_only;
   run;
ods excel close;
```

Figure 8-9 – The Excel 'Page Setup' Sheet Showing the Orientation set to 'LANDSCAPE'

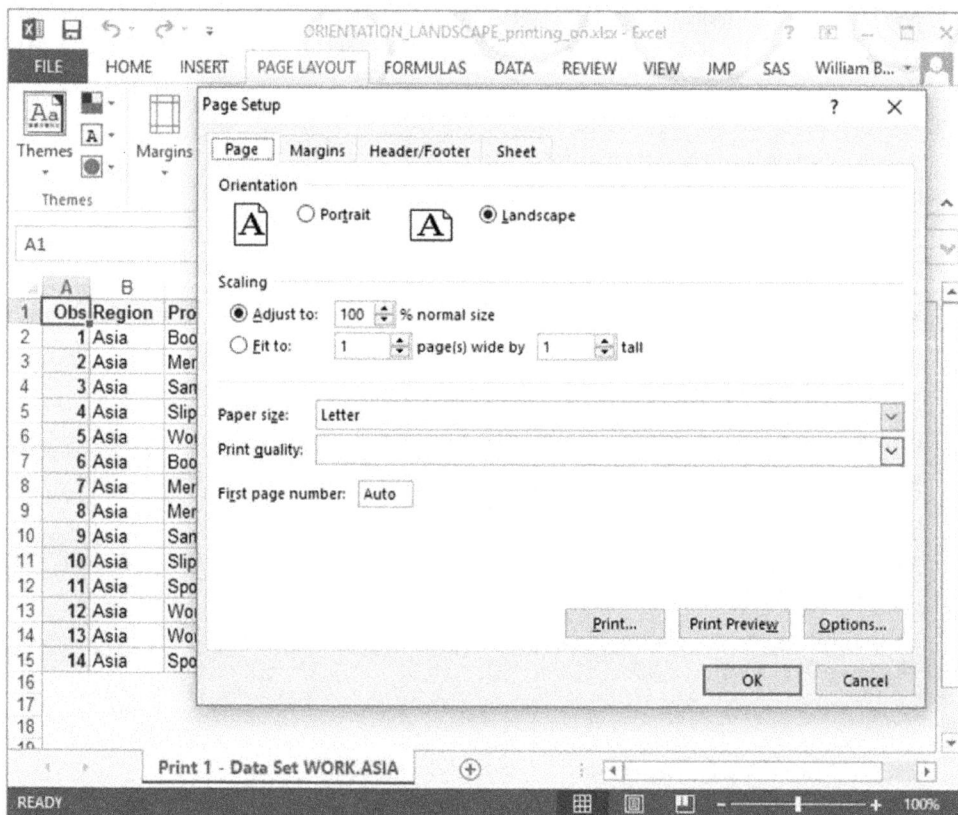

Page Setup 'LANDSCAPE' setting.

Figure 8-10 – Data Printed in 'LANDSCAPE' Mode

Obs	Region	Product	Subsidiary	Stores	Sales	Inventory	Returns
1	Asia	Boot	Bangkok	1	$1,996	$9,576	$80
2	Asia	Men's Dress	Bangkok	1	$3,033	$20,831	$52
3	Asia	Sandal	Bangkok	1	$3,230	$15,087	$120
4	Asia	Slipper	Bangkok	1	$3,019	$18,075	$127
5	Asia	Women's Casual	Bangkok	1	$5,389	$18,251	$185
6	Asia	Boot	Seoul	17	$60,712	$160,589	$1,296
7	Asia	Men's Casual	Seoul	1	$11,754	$2,176	$833
8	Asia	Men's Dress	Seoul	7	$116,333	$251,803	$2,443
9	Asia	Sandal	Seoul	3	$4,978	$21,483	$105
10	Asia	Slipper	Seoul	21	$149,013	$469,007	$2,941
11	Asia	Sport Shoe	Seoul	1	$937	$455	$10
12	Asia	Women's Casual	Seoul	2	$20,448	$38,576	$790
13	Asia	Women's Dress	Seoul	7	$78,234	$140,628	$1,891
14	Asia	Sport Shoe	Tokyo	1	$1,155	$15,602	$22

The SAS System

Full printed page in 'LANDSCAPE' mode to show actual location of the data on the printed page.

The PAGE_ORDER_ACROSS= Option

When your output from Excel does not print on one page, either because it is too wide or too long, Excel automatically chooses to print the data in sections. The default method that Excel uses is to print all data rows in the first group of columns that fit onto a page. That means that page 1 will have the data from the first *n* columns that fit on the page, and page 2 will have the data from same set of *n* columns for the following rows.

For example, suppose your data output has 20 columns and 100 rows. If 10 columns fit on a page and 50 rows fit on a page this would mean that columns 1 to 10 and rows 1 to 50 would be on page 1, and columns 1 to 10 and rows 51 to 100 would be on page 2. Page 3 would contain columns 11 to 20 and rows 1 to 50. The fourth and last page would contain columns 11 to 20 and rows 51 to 100. This is called page order "down, then over". If instead you want page 1 to have columns 1 to 10 and page 2 to have columns 11 to 20 you would use the page order "over, then down". Spreadsheets with more data would continue the pattern that you select for the printed output. It is of course your choice to determine the print order that you can read best, and then apply it to your output.

SAS Code 8-6 – Page Order Print Selection with PAGE_ORDER_ACROSS= Set to OFF and ON

```
ods excel (ID=1) file = "&path\PAGE_ORDER_ACROSS_printing_off.xlsx"
          options(PAGE_ORDER_ACROSS='Off' );
ods excel (ID=2) file = "&path\PAGE_ORDER_ACROSS_printing_on.xlsx"
          options(PAGE_ORDER_ACROSS='ON' );
   proc print data=Asia_only;
   run;
ods excel (ID=1) close;
ods excel (ID=2) close;
```

Figure 8-11 – Default Settings for Excel Page order, PAGE_ORDER_ACROSS='OFF'

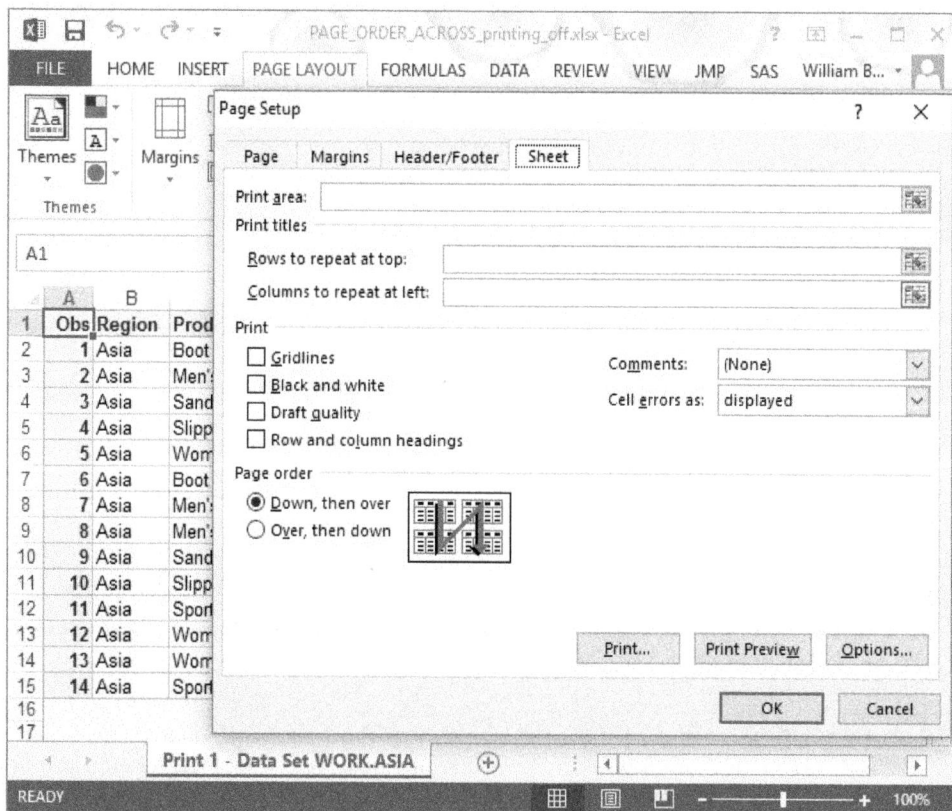

The arrow in the page order diagram goes down then to the next column. This is called "Down, then over."

Figure 8-12 – Settings for Excel Page Order, PAGE_ORDER_ACROSS='ON'

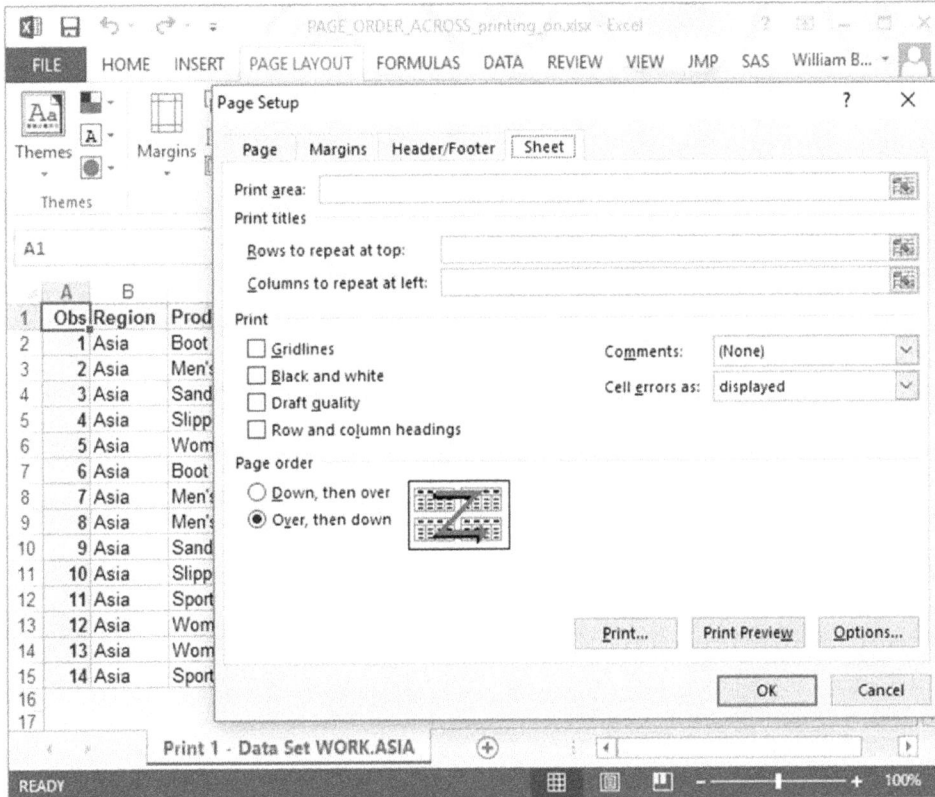

The arrow in the page order diagram goes over then down then to the set of rows. This is called "Over, then down."

The PRINT_AREA= Option

If you want to print only a select number of rows and columns this can be set up when your SAS program executes by using the ODS Excel option called PRINT_AREA=. Because any group of cells in the Excel spreadsheet always forms a rectangle (of course a square set of cells is also a rectangle), all you need to do is identify the left most top corner cell and the right most bottom cell. This does not change the spreadsheet image, but the codes that you supply appear in the "Print Area" of the page setup sheet. The default is that the whole worksheet is output, which is not shown.

SAS Code 8-7 – Code to Select a Small Print Area

```
ods excel file = "&path\PRINT_AREA.xlsx"
          options(PRINT_AREA='b,4,f,9' );
   proc print data=Asia_only;
   run;
ods excel close;
```

Figure 8-13 – The PRINT_AREA= Option to Print Only 30 Cells of the Spreadsheet

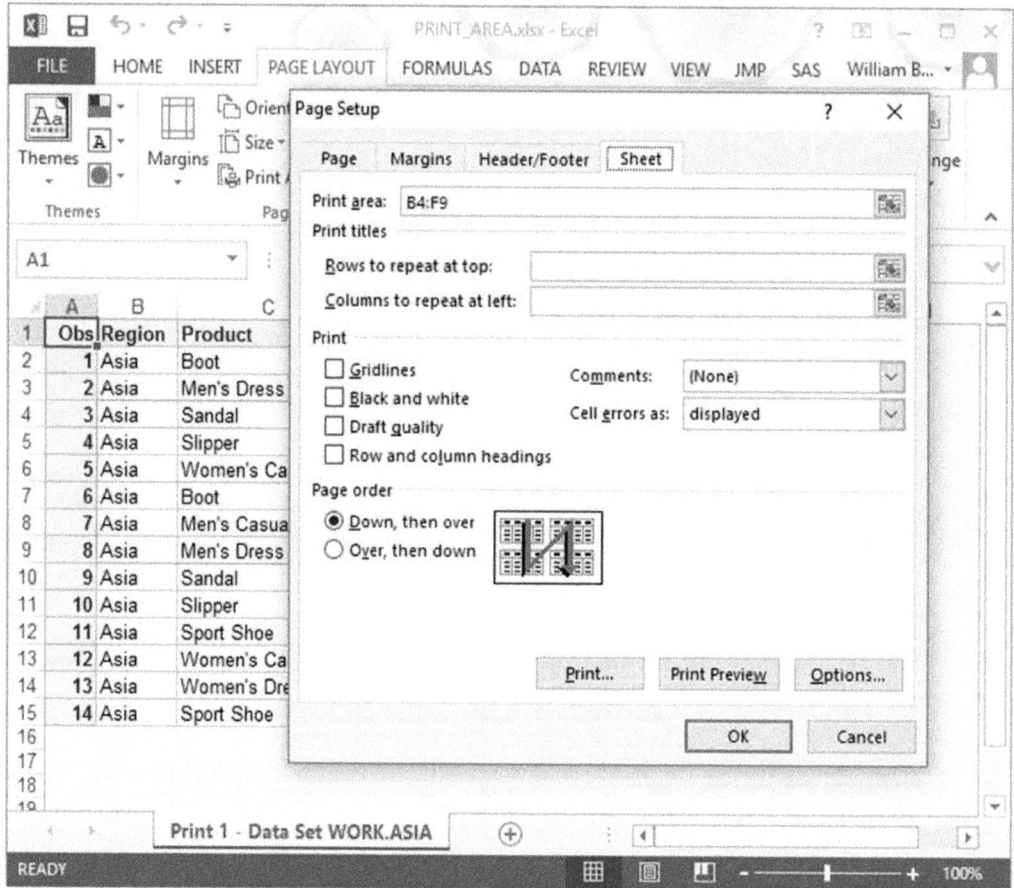

"Print area" shows the cell range of B4:F9; these cells will be the only ones printed.

Figure 8-14 – The PRINT_AREA= Output from SAS Code 8-7

		The SAS System		
Asia	Sandal	Bangkok	1	$3,230
Asia	Slipper	Bangkok	1	$3,019
Asia	Women's Casual	Bangkok	1	$5,389
Asia	Boot	Seoul	17	$60,712
Asia	Men's Casual	Seoul	1	$11,754
Asia	Men's Dress	Seoul	7	$116,333

This is the print preview output for this option. The page was printed in portrait mode and truncated to save space.

The PRINT_FOOTER= Option

Often companies want to have special company related or document-related information at the bottom of every printed page. The PRINT_FOOTER= option places the text string into the Excel worksheet area that places footers at the bottom of each page. This footer is often set manually, but SAS can insert the text at run time. A SAS footnote will override this option and will be used instead.

SAS Code 8-8 – Putting a Footer at the Bottom of Each Output Page

```
ods excel file = "&path\PRINT_FOOTER.xlsx"
          options(PRINT_FOOTER='My Footer Text' );
   proc print data=Asia_only;
   run;
ods excel close;
```

Figure 8-15 – Page Setup Sheet Showing the Excel Page Footer Information

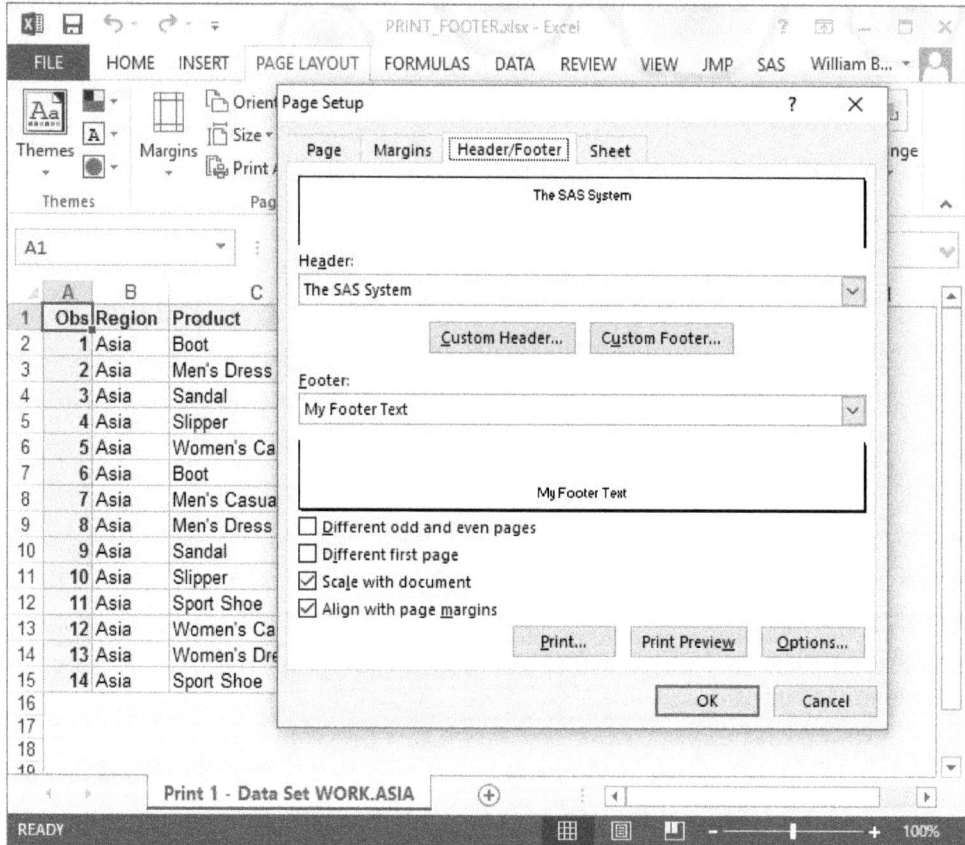

Remember that setting this footer using the SAS PRINT_FOOTER= option does not require you to open and change the output Excel workbook.

Figure 8-16 – Printed Page with the Excel Footer Displayed

The SAS System

Obs	Region	Product	Subsidiary	Stores	Sales	Inventory	Returns
1	Asia	Boot	Bangkok	1	$1,996	$9,576	$30
2	Asia	Men's Dress	Bangkok	1	$3,033	$20,831	$52
3	Asia	Sandal	Bangkok	1	$3,230	$15,087	$120
4	Asia	Slipper	Bangkok	1	$3,019	$16,075	$127
5	Asia	Women's Casual	Bangkok	1	$5,389	$16,251	$185
6	Asia	Boot	Seoul	17	$60,712	$160,589	$1,296
7	Asia	Men's Casual	Seoul	1	$11,754	$2,176	$833
8	Asia	Men's Dress	Seoul	7	$116,333	$251,803	$2,443
9	Asia	Sandal	Seoul	3	$4,978	$21,483	$105
10	Asia	Slipper	Seoul	21	$149,013	$469,007	$2,941
11	Asia	Sport Shoe	Seoul	1	$937	$455	$10
12	Asia	Women's Casual	Seoul	2	$20,448	$36,576	$790
13	Asia	Women's Dress	Seoul	7	$78,234	$140,628	$1,891
14	Asia	Sport Shoe	Tokyo	1	$1,155	$15,602	$22

My Footer Text

The full page is shown here to demonstrate the actual location of the text.

The PRINT_FOOTER_MARGIN= Option

When companies use preprinted paper for output documents, they might have an image at the bottom of the page that they want to preserve, without writing over the image. This can be accomplished by having a larger footer at the bottom of the page. Here, I will print a footer and use a three-inch footer margin to show this feature. The footer margin is measured in inches and the default value is 0.5 inches.

SAS Code 8-9 – The PRINT_FOOTER_MARGIN= Option

```
ods excel file = "&path\PRINT_FOOTER_MARGIN.xlsx"
          options(PRINT_FOOTER='My Footer Text'
                  PRINT_FOOTER_MARGIN='3' );
   proc print data=Asia_only;
   run;
ods excel close;
```

Figure 8-17 – Excel Page Setup Sheet Showing the Footer Set to 3 Inches

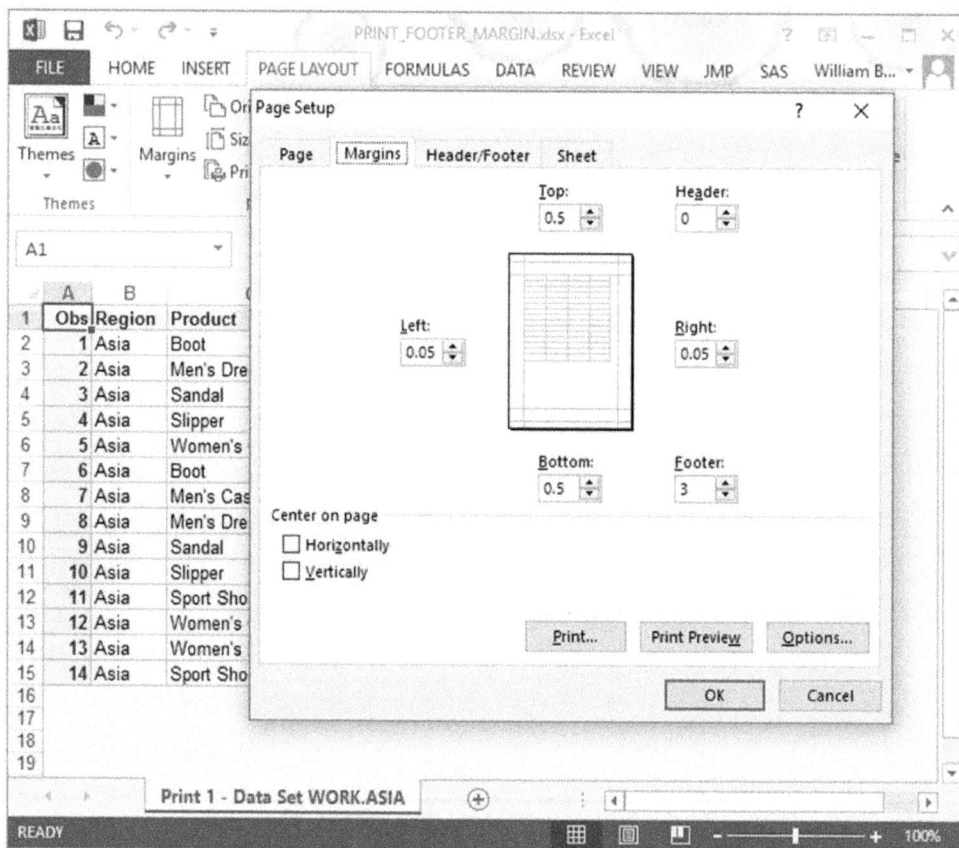

The footer box has a value of 3, and is measured in inches and reserves space at the bottom of the page.

Figure 8-18 –Applying the PRINT_FOOTER_MARGIN= Option

The SAS System

Obs	Region	Product	Subsidiary	Stores	Sales	Inventory	Returns
1	Asia	Boot	Bangkok	1	$1,996	$9,576	$80
2	Asia	Men's Dress	Bangkok	1	$3,033	$20,831	$52
3	Asia	Sandal	Bangkok	1	$3,230	$15,087	$120
4	Asia	Slipper	Bangkok	1	$3,019	$16,075	$127
5	Asia	Women's Casual	Bangkok	1	$5,389	$16,251	$185
6	Asia	Boot	Seoul	17	$60,712	$160,589	$1,296
7	Asia	Men's Casual	Seoul	1	$11,754	$2,176	$833
8	Asia	Men's Dress	Seoul	7	$116,333	$251,803	$2,443
9	Asia	Sandal	Seoul	3	$4,978	$21,483	$105
10	Asia	Slipper	Seoul	21	$149,013	$469,007	$2,941
11	Asia	Sport Shoe	Seoul	1	$937	$455	$10
12	Asia	Women's Casual	Seoul	2	$20,448	$36,576	$790
13	Asia	Women's Dress	Seoul	7	$78,234	$140,628	$1,891
14	Asia	Sport Shoe	Tokyo	1	$1,155	$15,602	$22

My Footer Text

This is the full page. Compare this to Figure 8-16 to see the difference between the default footer margin and the 3-inch margin.

The PRINT_HEADER= Option

Often companies want to have special company related or document-related information at the top of every printed page. The PRINT_HEADER= option places the text string into the Excel worksheet area that places headers at the top of each page. This header is often set manually, but SAS can insert the text at run time. A SAS title will override this option and will be used instead. To eliminate titles generated by SAS, like "The SAS System," use the TITLE statement.

SAS Code 8-10 – The PRINT_HEADER= Option

```
ods excel file = "&path\PRINT_HEADER.xlsx"
          options(PRINT_HEADER='My Header Text');
   title;
   proc print data=Asia_only;
   run;
ods excel close;
```

Figure 8-19 – The PRINT_HEADER= Page Setup Sheet

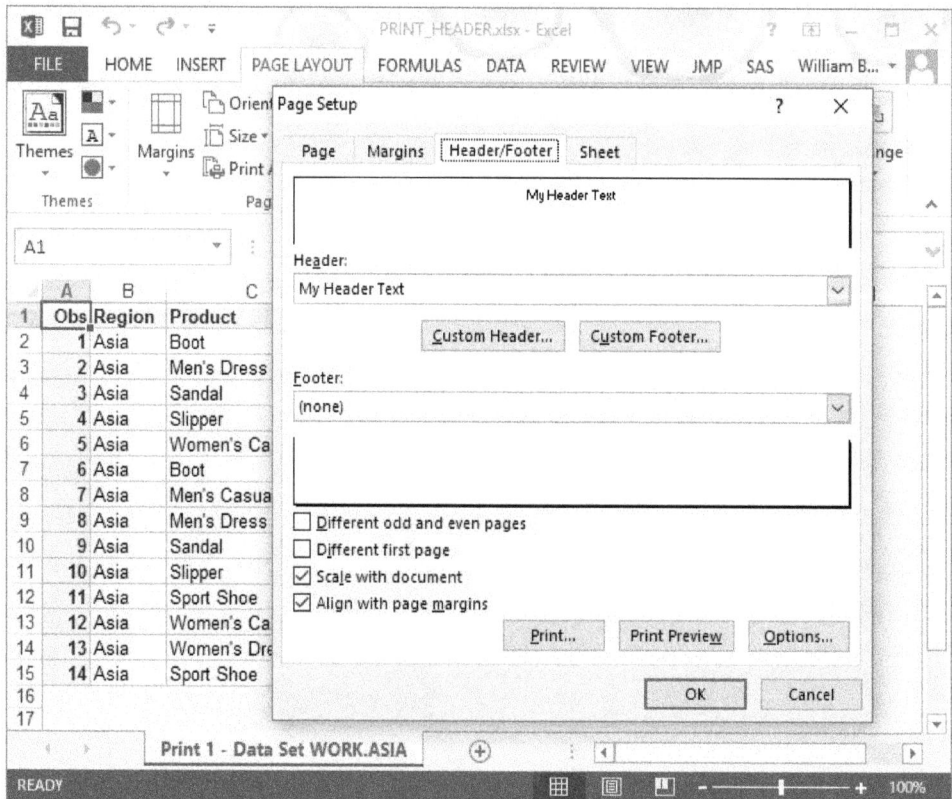

To get the text "My Header Text" at the top of the page in Figure 8-20 the SAS titles had to be turned off using the TITLE statement.

Figure 8-20 – The Printout Showing the Requested Text "My Header Text"

My Header Text

Obs	Region	Product	Subsidiary	Stores	Sales	Inventory	Returns
1	Asia	Boot	Bangkok	1	$1,996	$9,576	$80
2	Asia	Men's Dress	Bangkok	1	$3,033	$20,831	$52
3	Asia	Sandal	Bangkok	1	$3,230	$15,087	$120
4	Asia	Slipper	Bangkok	1	$3,019	$16,075	$127
5	Asia	Women's Casual	Bangkok	1	$5,389	$16,251	$185
6	Asia	Boot	Seoul	17	$60,712	$160,589	$1,296
7	Asia	Men's Casual	Seoul	1	$11,754	$2,176	$833
8	Asia	Men's Dress	Seoul	7	$116,333	$251,803	$2,443
9	Asia	Sandal	Seoul	3	$4,978	$21,483	$105
10	Asia	Slipper	Seoul	21	$149,013	$469,007	$2,941
11	Asia	Sport Shoe	Seoul	1	$937	$455	$10
12	Asia	Women's Casual	Seoul	2	$20,448	$36,576	$790
13	Asia	Women's Dress	Seoul	7	$78,234	$140,628	$1,891
14	Asia	Sport Shoe	Tokyo	1	$1,155	$15,602	$22

The full printed page showing the positioning of the output header text.

The PRINT_HEADER_MARGIN= Option

The Excel "Header" setting on the Page Setup sheet seems to act independently from the "Top" margin setting on the Page Setup sheet. I need to spend a little more time here to explain how it works. My example SAS Code 8-11 shows several options, not just the PRINT_HEADER_MARGIN= option. What this example does is output three Excel workbooks. I will show one Page Setup sheet and describe how the three different settings for the PRINT_HEADER_MARGIN= setting affect the output printout. In this example, I will also show the margin and heading tabs on the preview sheet to help explain the examples. The default header setting is 0.0 (zero) inches, which is also the common default for the top margin setting. I also reduced the amount of PRINT_AREA= printed output and added a PRINT_HEADER= text string. In this way I can show that both are output, but the header is hidden by the information from the printed output Excel data cells.

SAS Code 8-11 – Examples of Using the PRINT_HEADER_MARGIN Settings

```
ods excel (id=1) file = "&path\PRINT_HEADER_Margin_0_5.xlsx"
          options(PRINT_HEADER='My Header Text'
                  PRINT_HEADER_MARGIN='0.5'
                  PRINT_AREA='b,4,f,9');
ods excel (id=2) file = "&path\PRINT_HEADER_Margin_1.xlsx"
          options(PRINT_HEADER='My Header Text'
                  PRINT_HEADER_MARGIN='1'
                  PRINT_AREA='b,4,f,9');
ods excel (id=3) file = "&path\PRINT_HEADER_Margin_4.xlsx"
          options(PRINT_HEADER='My Header Text'
                  PRINT_HEADER_MARGIN='4'
                  PRINT_AREA='b,4,f,9');
   title ;
   proc print data=Asia_only;
   run;

ods excel (id=1) close;
ods excel (id=2) close;
ods excel (id=3) close;
```

Figure 8-21 – Excel Page Setup Sheet Showing the Header Set to 0.5 Inches

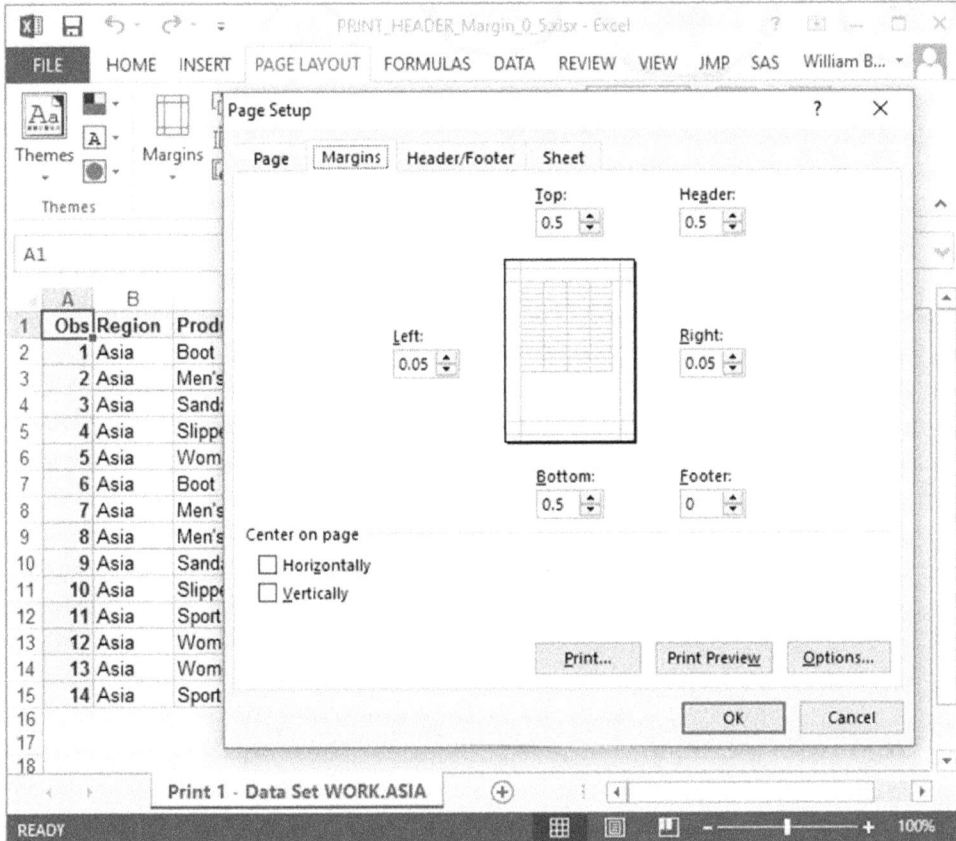

This screen shot shows the output of the first step in the SAS Code 8-11 example where the PRINT_HEADER_MARGIN= value is set to 0.5 inches. The other output files (ID=2 and ID=3), which are not shown here, set the value in the header box to 1 and 4 inches respectively. The print preview pages are shown in Figures 8-22, 8-23, and 8-24.

Figure 8-22 – PRINT_HEADER_MARGIN=0.5 and Excel Top Margin=0.5

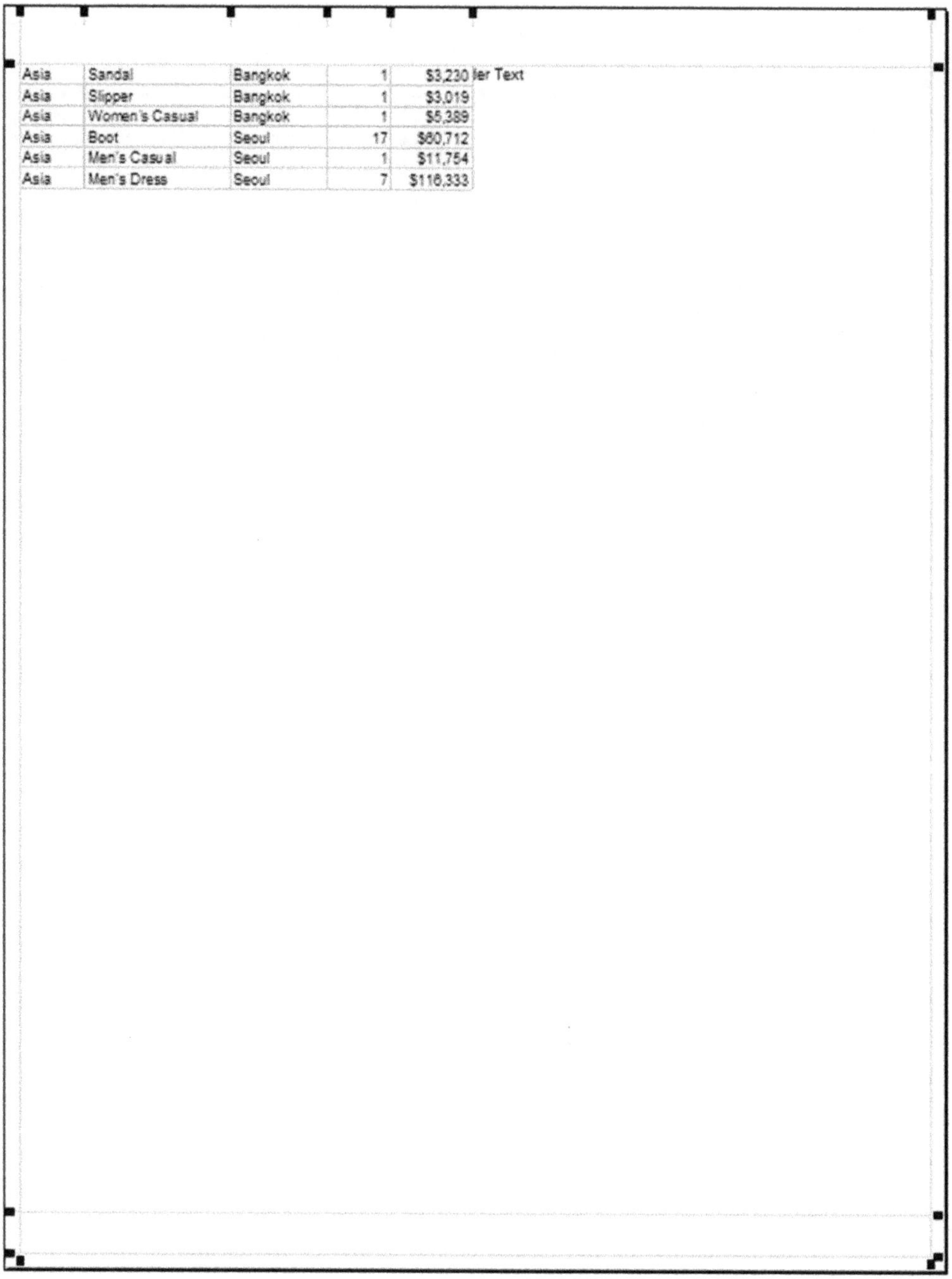

Asia	Sandal	Bangkok	1	$3,230	er Text
Asia	Slipper	Bangkok	1	$3,019	
Asia	Women's Casual	Bangkok	1	$5,389	
Asia	Boot	Seoul	17	$60,712	
Asia	Men's Casual	Seoul	1	$11,754	
Asia	Men's Dress	Seoul	7	$116,333	

Printed output with the default top margin of 0.5 inches and the PRINT_HEADER_MARGIN= of 0.5 inches. This shows that both the default Excel top margin and the ODS PRINT_HEADER_MARGIN= are aligned.

Figure 8-23 – PRINT_HEADER_MARGIN=1.0 and Excel Top margin=0.5

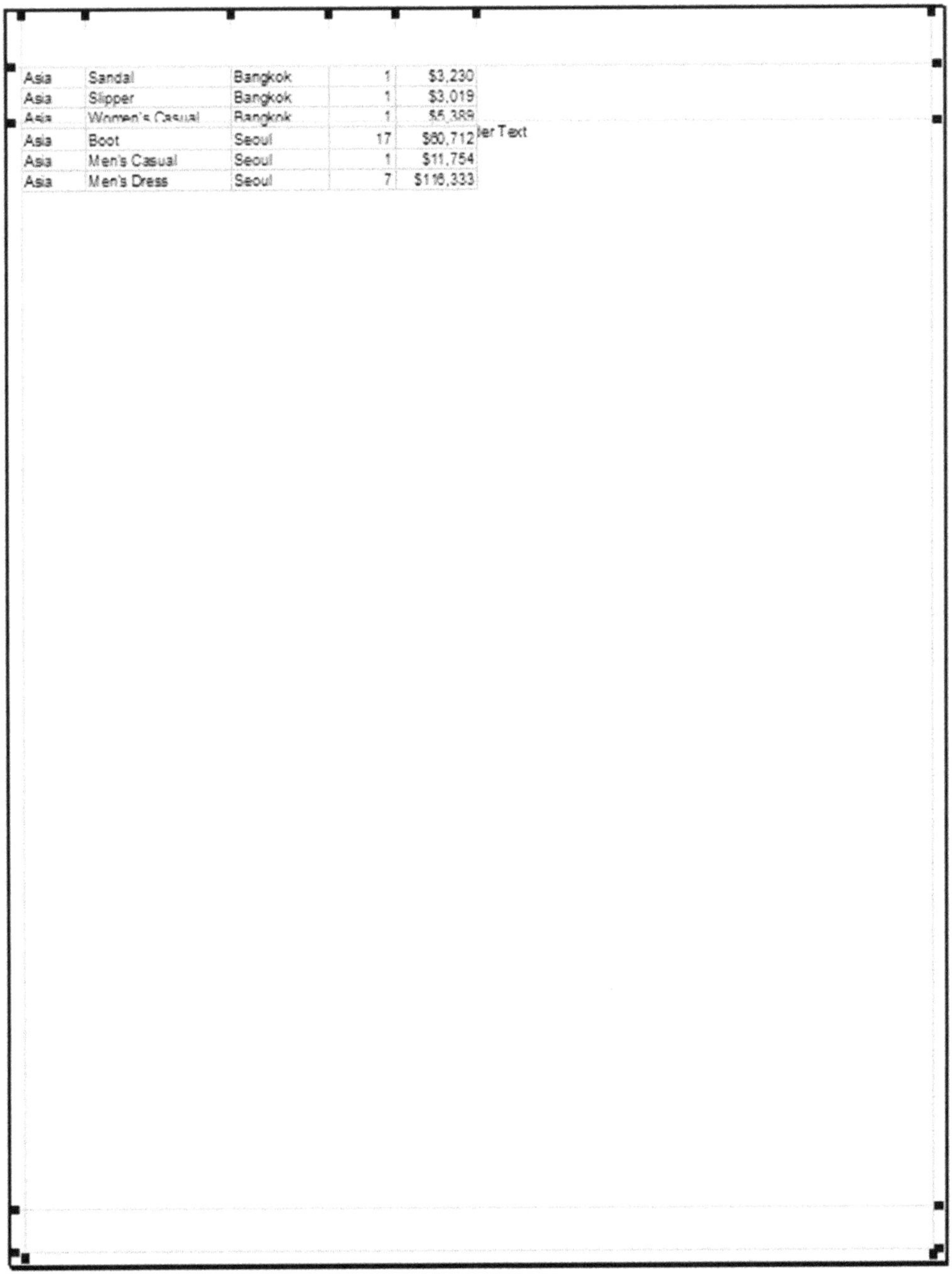

Asia	Sandal	Bangkok	1	$3,230
Asia	Slipper	Bangkok	1	$3,019
Asia	Women's Casual	Bangkok	1	$5,389
Asia	Boot	Seoul	17	$60,712
Asia	Men's Casual	Seoul	1	$11,754
Asia	Men's Dress	Seoul	7	$116,333

Printed output with the default top margin of 0.5 inches and the PRINT_HEADER_MARGIN= of 1.0 inches. This shows that the default Excel top margin and the ODS PRINT_HEADER_MARGIN are applied separately.

Figure 8-24 – PRINT_HEADER_MARGIN=4.0 and Excel Top Margin=0.5

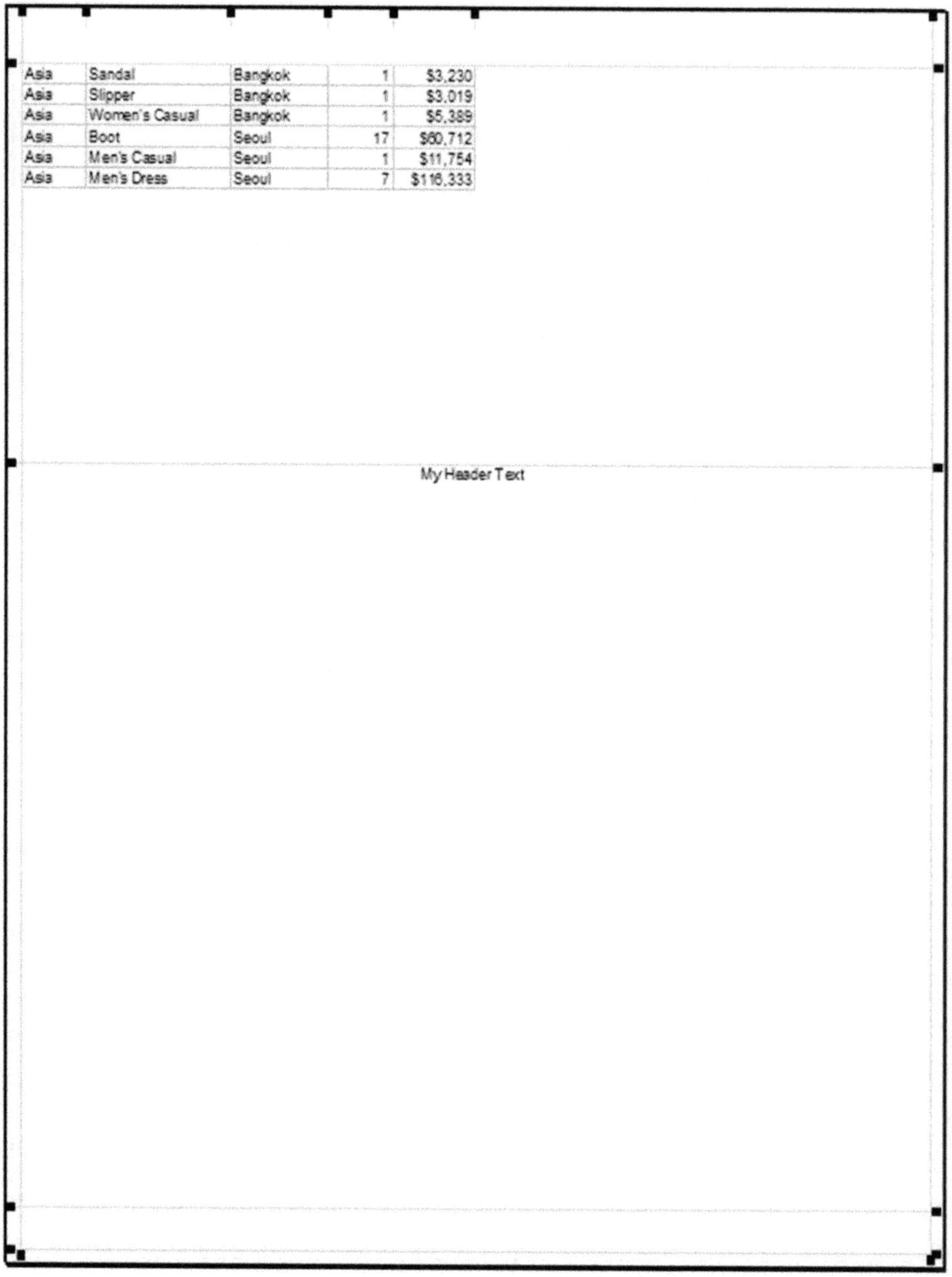

Asia	Sandal	Bangkok	1	$3,230
Asia	Slipper	Bangkok	1	$3,019
Asia	Women's Casual	Bangkok	1	$5,389
Asia	Boot	Seoul	17	$60,712
Asia	Men's Casual	Seoul	1	$11,754
Asia	Men's Dress	Seoul	7	$116,333

My Header Text

Printed output with the default top margin of 0.5 inches and the PRINT_HEADER_MARGIN= of 4.0 inches. This shows that the default Excel top margin and the ODS PRINT_HEADER_MARGIN= are applied separately, and that the PRINT_HEADER_MARGIN= is also independent of the printed data.

The ROWBREAKS_COUNT= Option

The ROWBREAKS_COUNT= option allows you as a programmer to break the output pages into groups the size that you want. In SAS Code 8-12, I have used the value of '10'. The results of this option are not always easy to see without printing the output pages. However, if you call up the print preview and then go back to the standard sheet display new dotted lines are displayed. These dotted lines represent the page breaks that are used when the page is printed.

SAS Code 8-12 – ROWBREAKS_COUNT= Set to a Value of 10

```
ods excel file = "&path\ROWBREAKS_COUNT.xlsx"
          options(ROWBREAKS_COUNT='10');
   proc print data=Asia_only;
   run;
ods excel close;
```

Figure 8-25 –Applying the ROWBREAKS_COUNT=10 Option.

	Obs	Region	Product	Subsidiary	Stores	Sales	Inventory	Returns
1								
2	1	Asia	Boot	Bangkok	1	$1,996	$9,576	$80
3	2	Asia	Men's Dress	Bangkok	1	$3,033	$20,831	$52
4	3	Asia	Sandal	Bangkok	1	$3,230	$15,087	$120
5	4	Asia	Slipper	Bangkok	1	$3,019	$16,075	$127
6	5	Asia	Women's Casual	Bangkok	1	$5,389	$16,251	$185
7	6	Asia	Boot	Seoul	17	$60,712	$160,589	$1,296
8	7	Asia	Men's Casual	Seoul	1	$11,754	$2,176	$833
9	8	Asia	Men's Dress	Seoul	7	$116,333	$251,803	$2,443
10	9	Asia	Sandal	Seoul	3	$4,978	$21,483	$105
11	10	Asia	Slipper	Seoul	21	$149,013	$469,007	$2,941
12	11	Asia	Sport Shoe	Seoul	1	$937	$455	$10
13	12	Asia	Women's Casual	Seoul	2	$20,448	$36,576	$790
14	13	Asia	Women's Dress	Seoul	7	$78,234	$140,628	$1,891
15	14	Asia	Sport Shoe	Tokyo	1	$1,155	$15,602	$22

The vertical dotted line and the line below "Obs 10" represent the page breaks that will occur. Also notice that the ROWBREAKS_COUNT= splits the data after data lines, and the first row of headers is not counted as one of the 10 lines. You have to use other options to reproduce the row headers on the second and following pages.

The ROWBREAKS_INTERVAL= Option

The ROWBREAKS_INTERVAL= option works best when SHEET_INTERVAL='NONE' is selected. When you choose not to place all information onto one sheet, then the breaks most likely will cause a new Excel worksheet to be generated as opposed to a new printed page. The example shown in SAS Code 8-13 writes three workbooks with the same sorted data set printed twice by the 'PRODUCT' values. The final Excel output workbook also has the scale setting set to 50% to fit everything onto one page. Figures 8-26 to 8-28 show only the full first page of the Excel Print Preview screen. Shown at the bottom of Figure 8-26 is the page counter. I do not show it here, but all of the data for each of these examples is on one Excel worksheet.

SAS Code 8-13 – The ROWBREAKS_INTERVAL= Options Shown for All Options

```
ods excel (id=1)file = "&path\ROWBREAKS_INTERVAL_out.xlsx"
          options(ROWBREAKS_INTERVAL='output'
                  sheet_interval="none");
ods excel (id=2)file = "&path\ROWBREAKS_INTERVAL_proc.xlsx"
          options(ROWBREAKS_INTERVAL='proc'
                  sheet_interval="none");
ods excel (id=3)file = "&path\ROWBREAKS_INTERVAL_none.xlsx"
          options(ROWBREAKS_INTERVAL='none'
                  sheet_interval="none"
                  scale='50');
 proc sort data=Asia_only out=temp;
  by product;
  run;
  proc print data=temp;
 by product;
  run;
  proc print data=temp;
 by product;
  run;
ods excel (id=1) close;
ods excel (id=2) close;
ods excel (id=3) close;
```

Figure 8-26 – ROWBREAKS_INTERVAL= Set to Page for Each Output Change

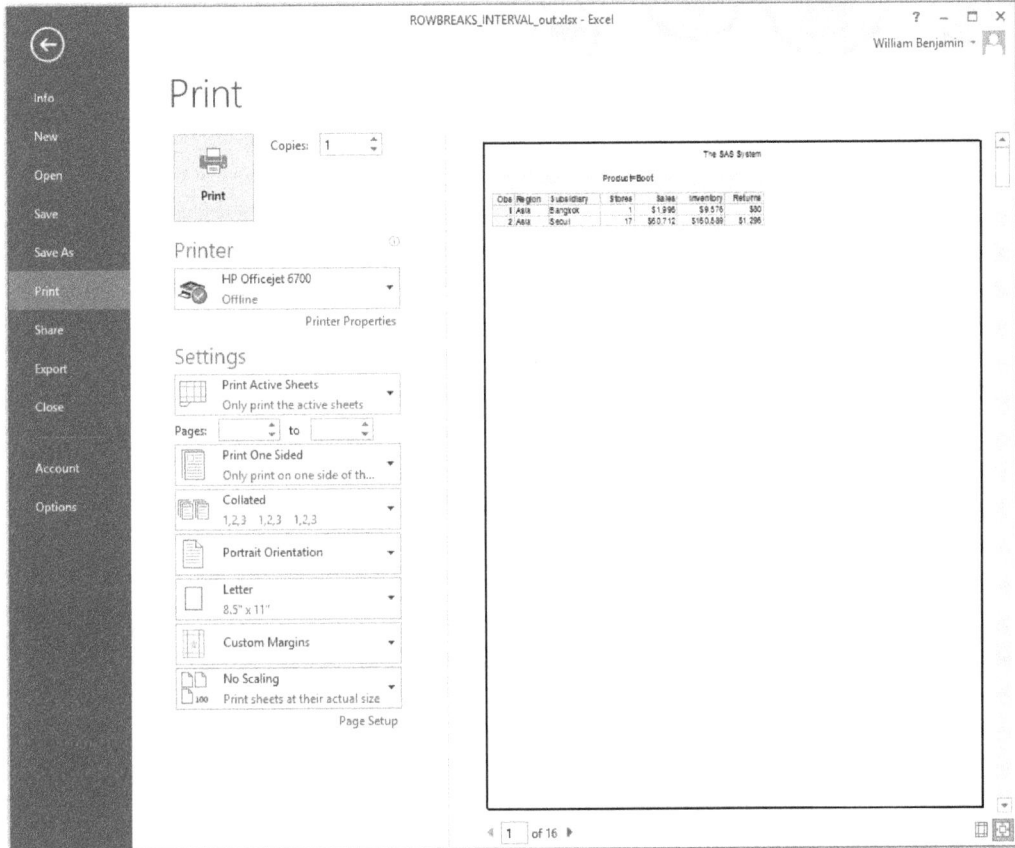

All the information for the examples in Figures 8-26 to 8-28 were output to one Excel worksheet, but the data is printed on a different number of pages for each example. In Figure 8-26 the data is split by each output unit and is printed on 16 pages.

Figure 8-27 – ROWBREAKS_INTERVAL= Set to Page for Each Procedure Change

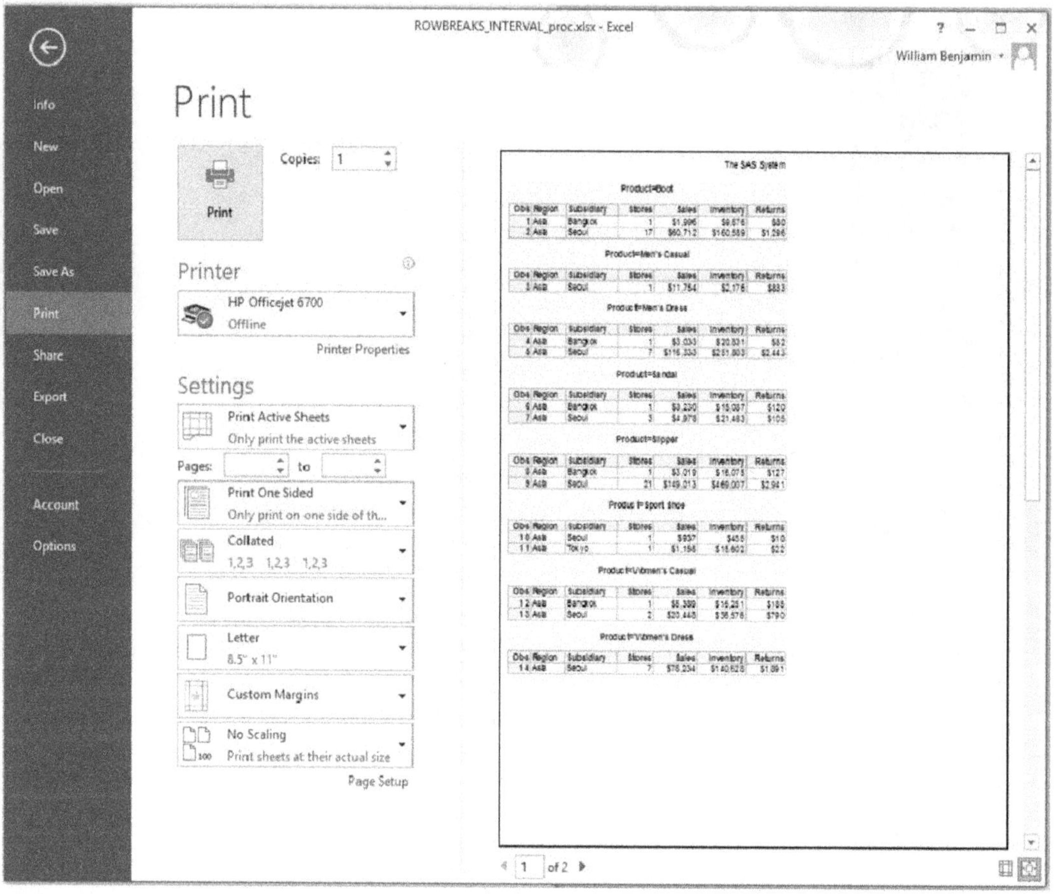

In Figure 8-27 the data is split by each procedure unit and is printed on two pages.

Figure 8-28 – ROWBREAKS_INTERVAL= Set to None So Only One Page Is Output

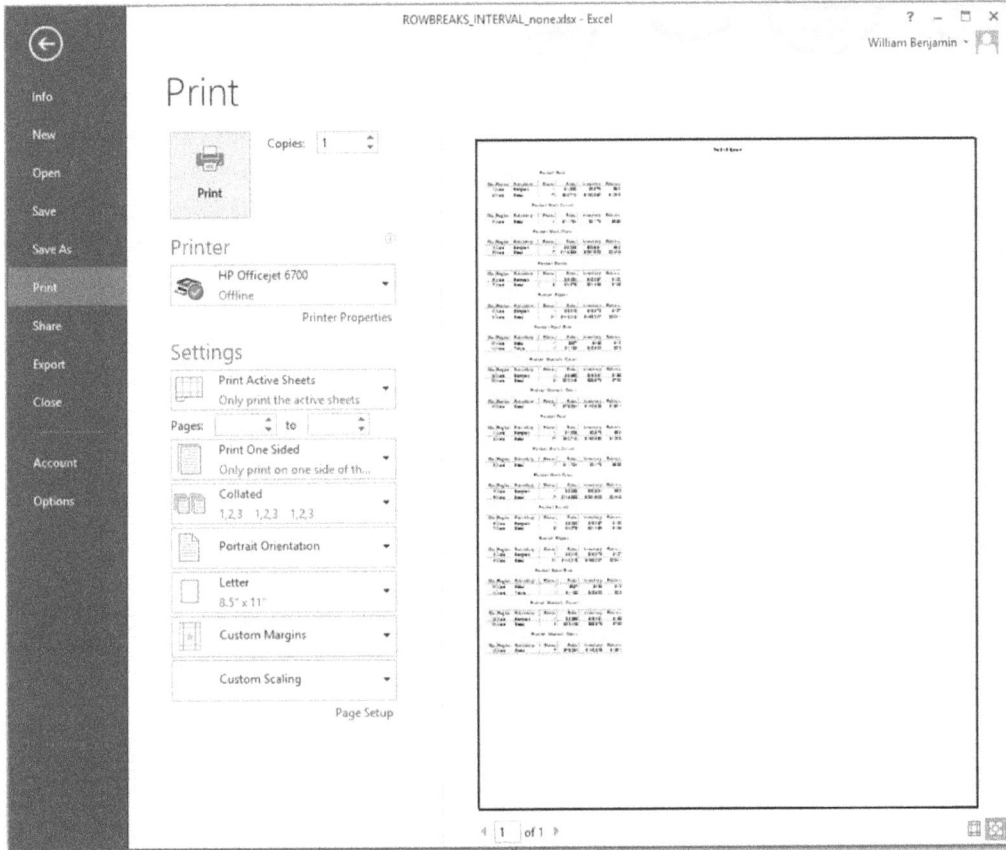

In Figure 8-28 the data is not split and is printed on one page.

The ROWCOLHEADINGS= Option

This option controls whether or not the printed Excel output has the Excel "ROW" and "COLUMN" values printed. The default is not to print these Excel headings. The Excel "ROW" headings are the row numbers on the left side of the printed spreadsheet, and the "COLUMN" headings are the letters at the top of the printed spreadsheet. Selecting "ON" turns the printing on. The default option is "OFF".

SAS Code 8-14 – The ODS Excel Destination ROWCOLHEADINGS= Option Turned On

```
ods excel file = "&path\ROWCOLHEADINGS_on.xlsx"
          options(ROWCOLHEADINGS='on');
   proc print data=Asia_only;
   run;
ods excel close;
```

Figure 8-29 – Excel Page Setup Sheet Showing the Check Box with the Row and Column Headings Selected

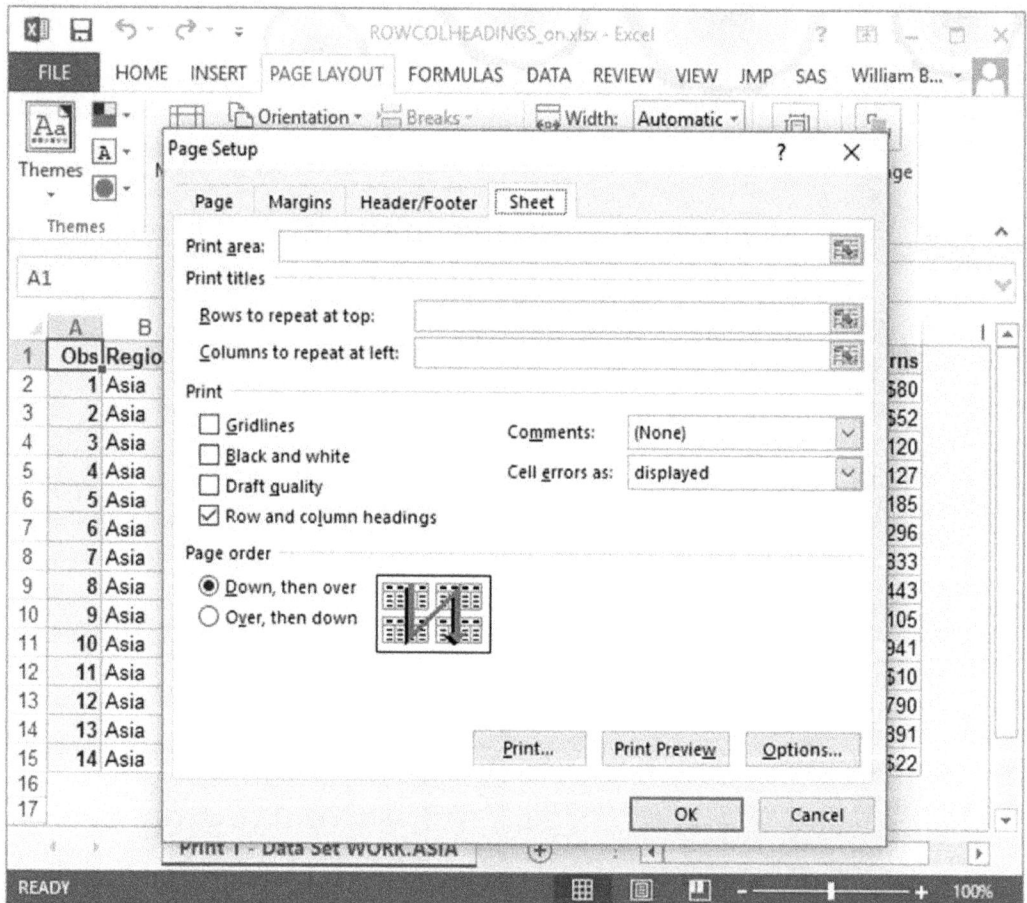

This image shows the "Row and column headings" check box selected indicating that Excel "Row" and "Column" headers should be printed.

Figure 8-30 – Excel Print Page Showing the Excel "Row" and "Column" Headers

The SAS System

	A	B	C	D	E	F	G	H
1	Obs	Region	Product	Subsidiary	Stores	Sales	Inventory	Returns
2	1	Asia	Boot	Bangkok	1	$1,996	$9,576	$80
3	2	Asia	Men's Dress	Bangkok	1	$3,033	$20,831	$52
4	3	Asia	Sandal	Bangkok	1	$3,230	$15,087	$120
5	4	Asia	Slipper	Bangkok	1	$3,019	$16,075	$127
6	5	Asia	Women's Casual	Bangkok	1	$5,389	$16,251	$185
7	6	Asia	Boot	Seoul	17	$60,712	$160,589	$1,296
8	7	Asia	Men's Casual	Seoul	1	$11,754	$2,176	$833
9	8	Asia	Men's Dress	Seoul	7	$116,333	$251,803	$2,443
10	9	Asia	Sandal	Seoul	3	$4,978	$21,483	$105
11	10	Asia	Slipper	Seoul	21	$149,013	$469,007	$2,941
12	11	Asia	Sport Shoe	Seoul	1	$937	$455	$10
13	12	Asia	Women's Casual	Seoul	2	$20,448	$36,576	$790
14	13	Asia	Women's Dress	Seoul	7	$78,234	$140,628	$1,891
15	14	Asia	Sport Shoe	Tokyo	1	$1,155	$15,602	$22
16								

This page was output as a portrait formatted page, but it is truncated to save space.

The SCALE= Option

Often I have found it hard to get a full set of variables on one printed page, The ODS Excel SCALE= option helps me reduce the output to allow more on a page. The example below in SAS Code 8-15 leaves the portrait printing alone, but it reduces the size of the printed material. The default is to scale the output at 100%. To set another value, place a positive number within quotation marks, as shown in SAS Code 8-15. I took some time to play with this option over the years and found some interesting things. I am currently using Excel 2013, and the range for the Excel SCALE= option is 10% to 400% normal size. If you enter a nonnumeric value into the string in your SAS code, SAS passes 100 to Excel. SAS does not check the value that you enter, only that it is numeric. I have noticed the following results with Excel 2013. I do not expect differences but they might occur in other versions of Excel.

When Using Excel 2013, If You Enter:

- A nonnumeric value – SAS does not generate an error and SAS passes 100 to Excel.
- A negative number – SAS does not generate an error and Excel uses 100%.
- The number zero – SAS does not generate an error and Excel uses 10%.
- Numbers from 1 to 9 – SAS does not generate an error and Excel uses "AUTO" as the "PAGE LAYOUT" SCALE value. On the "Page Setup" page an error message is shown when you try to use the "PAGE" tab. The message says "The number must be between

10 and 400. Try again by entering a number in this range." Then you have to cancel and reenter a number in the "SCALE" instead of "AUTO."

- 10 to 400 = ok

- A number between 401 and 32767 – SAS does not generate an error message and Excel uses the number entered as the "PAGE LAYOUT" SCALE value and the "Page Setup" scale value on the "PAGE" tab. No Excel error is generated and you can proceed to the "Print Preview" page. However, when you exit the "Print Preview" page the scale value reverts to 400 in both locations. Also, any attempt to use the toggle arrows to increase or decrease the scale values causes the value to revert to 400%.

- Numbers over 32767 ((2 ** 15) -1) cause Excel to generate an error when accessing the "PAGE" tab on the "Page Setup" sheet. This value is commonly referred to as 32K because when starting at zero exactly 2 to the 15th power numbers exist. In binary this would range from '000000000000000'b to '111111111111111'b.

SAS Code 8-15 – The ODS Excel SCALE Option Set to 50% of Normal Size

```
ods excel file = "&path\SCALE_50.xlsx"
          options(SCALE='50');
  proc print data=Asia_only;
  run;
ods excel close;
```

Figure 8-31 – Result of the ODS Excel Destination "SCALE=" Option When Set to '50'

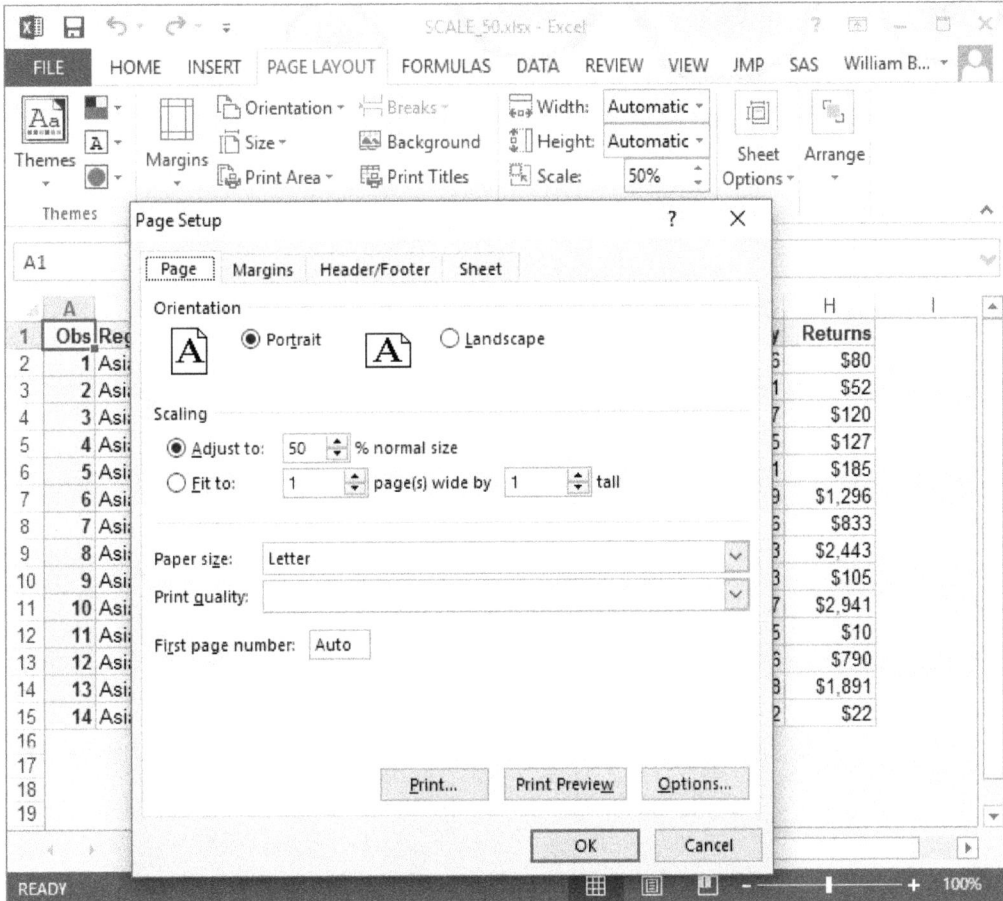

You can view the "SCALE" value both on the Excel "Page Layout" tab and on the "Page Setup" sheet shown here.

Figure 8-32 – The Output When "SCALE='50'" Is Selected

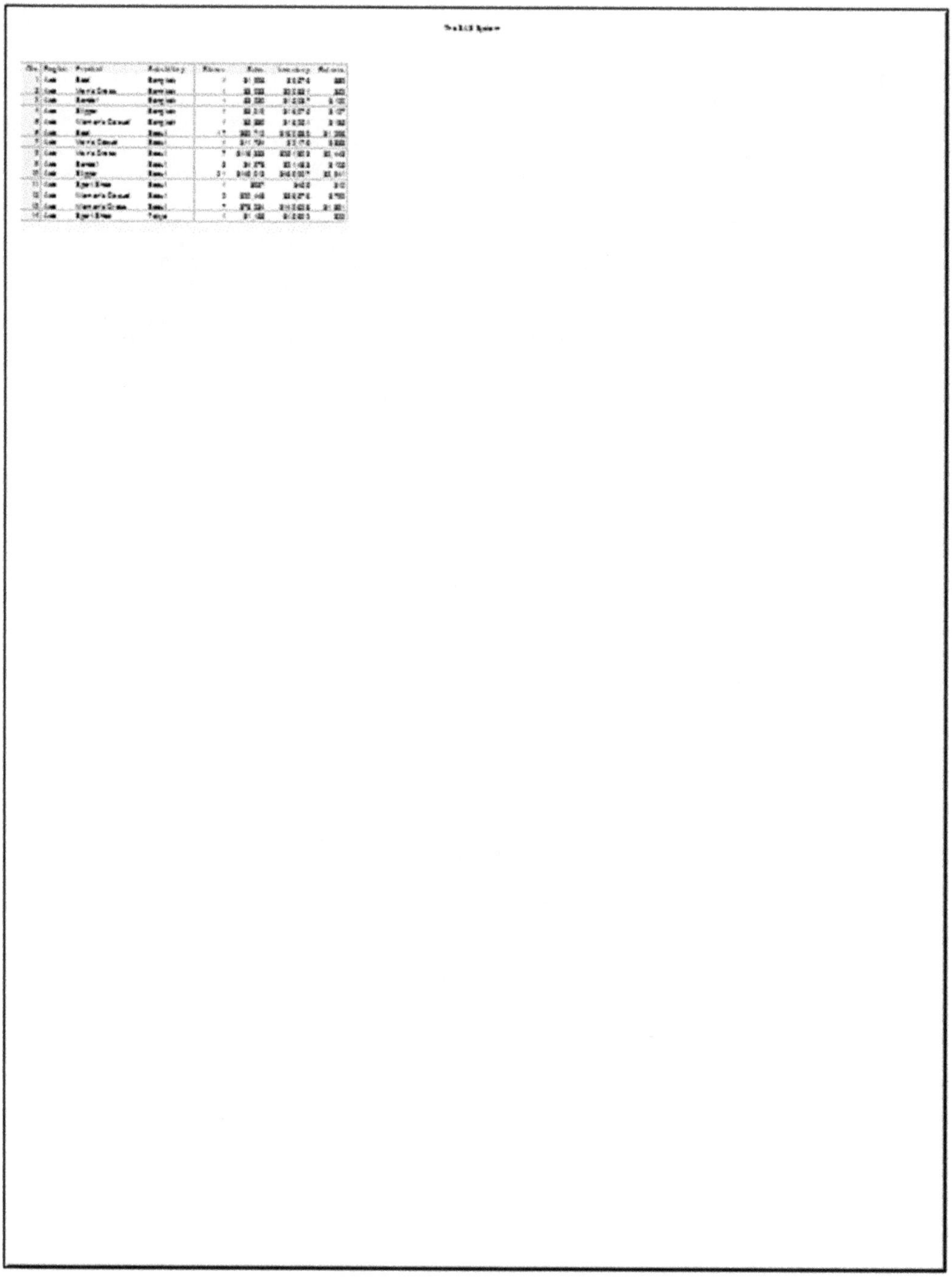

This is a full page of the printed output. Yes, at this size it is nearly unreadable especially if you are reading this on a small screen. But the decreased size of the fonts does enable you to place more information on a sheet. This works better if you are printing in Landscape mode and have more columns.

The TITLE_FOOTNOTE_NOBREAK= Option

The ODS Excel destination TITLE_FOOTNOTE_NOBREAK= option is a tricky one. This is one of those "Yes, we have no bananas today" type of options. Many people would call this a double negative option. The "NOBREAK" part of the option says do not break titles and footnotes in the output Excel workbook. The option "YES" says do not break the titles and footnotes. The default of not using the option and the "NO" value of the option negates the "NOBREAK" part of the option name and means to wrap the titles and footnotes if they are long. Yes, I know that was confusing. Hopefully the code and screen shots below will help. I will show the output directly below the code segments that produce the output Excel workbooks.

SAS Code 8-16 – ODS Excel TITLE_FOOTNOTE_NOBREAK= Option (Default Setting)

```
title    "This a my long title so I can show that it wraps around
when title_footnote_nobreak is not active.";
footnote "This a my very long footnote (it prints in a smaller font)
so I can show that it wraps around when title_footnote_nobreak not
active.";

ods excel file = "&path\TITLE_FOOTNOTE_NOBREAK_default.xlsx"
         options(embedded_titles='yes'
                 embedded_footnotes='yes');
 proc print data=asia_only;
   run;
ods excel close;
```

The SAS Code 8-16 code is the default setting for the TITLE_FOOTNOTE_NOBREAK= option and causes the titles and footnotes to wrap to the next line if the title or footnote is too long.

Figure 8-33 – The TITLE_FOOTNOTE_NOBREAK= Option (Default Setting)

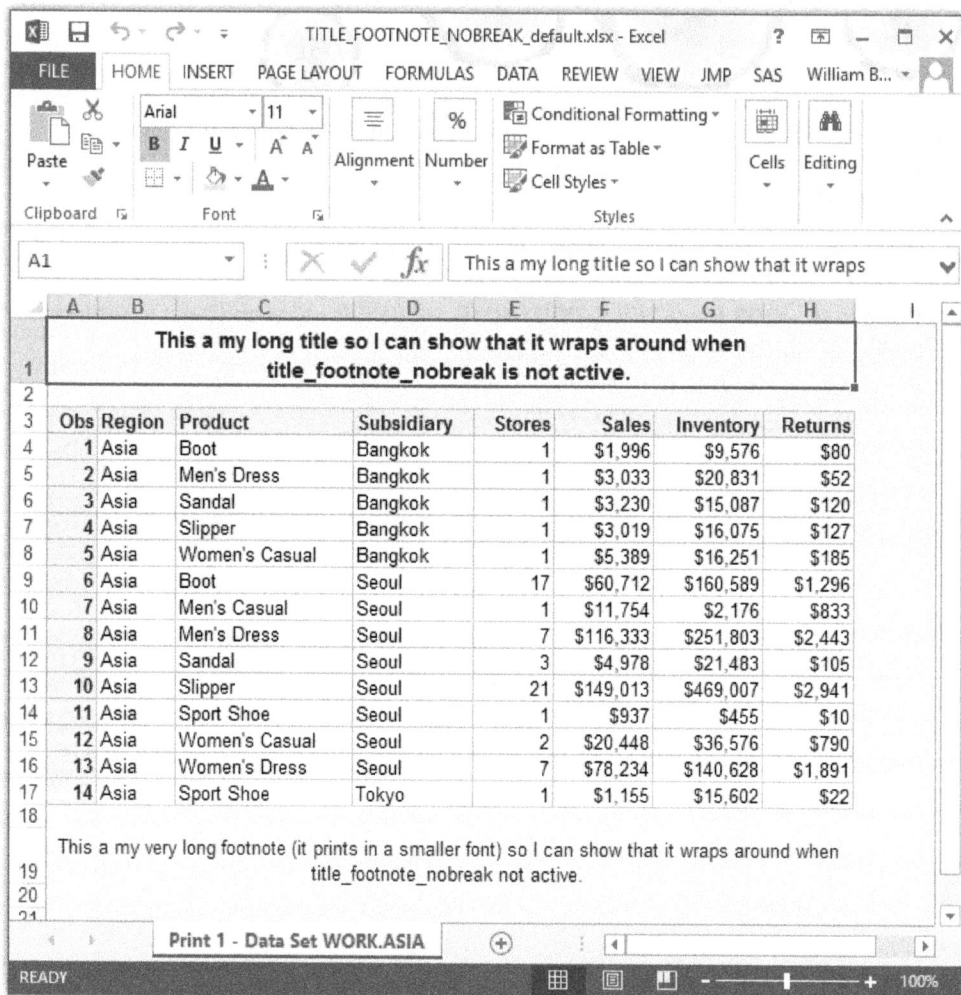

Without using the ODS EXCEL TITLE_FOOTNOTE_NOBREAK= option the titles and footnotes wrap to the next line if the title or footnote is too long. Also, note that the Titles and footnotes are printed in different font sizes and have different bold options.

The code in SAS CODE 8-17 sets the TITLE_FOOTNOTE_NOBREAK= option to 'NO', which produces the same result as the default option.

SAS Code 8-17 – ODS Excel TITLE_FOOTNOTE_NOBREAK= Option Set to "NO"

```
title    "This a my long title so I can show that it wraps around
when title_footnote_nobreak=''no''";

footnote "This a my very long footnote (it prints in a smaller font)
so I can show that it wraps around when
title_footnote_nobreak=''no''";
ods excel file = "&path\TITLE_FOOTNOTE_NOBREAK_no.xlsx"
         options(title_footnote_nobreak='no'
```

```
                    embedded_titles='yes'
                    embedded_footnotes='yes');
 proc print data=asia_only;
  run;
ods excel close;
```

SAS Code 8-17 code is set to NO for the TITLE_FOOTNOTE_NOBREAK= option and it causes the titles and footnotes to wrap to the next line if the title or footnote is too long.

Figure 8-34 – The TITLE_FOOTNOTE_NOBREAK= Option Set to "NO"

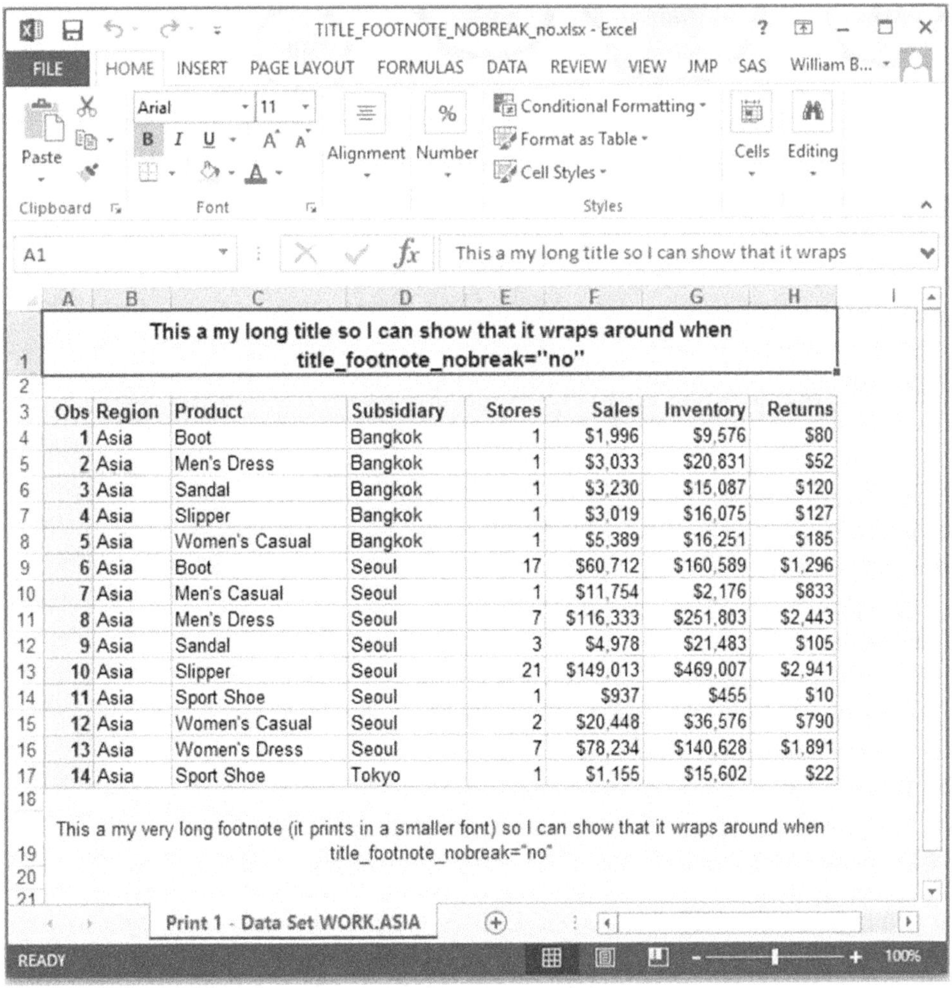

This a my long title so I can show that it wraps around when title_footnote_nobreak="no"

Obs	Region	Product	Subsidiary	Stores	Sales	Inventory	Returns
1	Asia	Boot	Bangkok	1	$1,996	$9,576	$80
2	Asia	Men's Dress	Bangkok	1	$3,033	$20,831	$52
3	Asia	Sandal	Bangkok	1	$3,230	$15,087	$120
4	Asia	Slipper	Bangkok	1	$3,019	$16,075	$127
5	Asia	Women's Casual	Bangkok	1	$5,389	$16,251	$185
6	Asia	Boot	Seoul	17	$60,712	$160,589	$1,296
7	Asia	Men's Casual	Seoul	1	$11,754	$2,176	$833
8	Asia	Men's Dress	Seoul	7	$116,333	$251,803	$2,443
9	Asia	Sandal	Seoul	3	$4,978	$21,483	$105
10	Asia	Slipper	Seoul	21	$149,013	$469,007	$2,941
11	Asia	Sport Shoe	Seoul	1	$937	$455	$10
12	Asia	Women's Casual	Seoul	2	$20,448	$36,576	$790
13	Asia	Women's Dress	Seoul	7	$78,234	$140,628	$1,891
14	Asia	Sport Shoe	Tokyo	1	$1,155	$15,602	$22

This a my very long footnote (it prints in a smaller font) so I can show that it wraps around when title_footnote_nobreak="no"

When using the ODS EXCEL TITLE_FOOTNOTE_NOBREAK= option set to NO, the titles and footnotes wrap to the next line if the title or footnote is too long. Also, note that the titles and footnotes are printed in different font sizes and have different bold options.

The code in SAS CODE 8-18 sets the TITLE_FOOTNOTE_NOBREAK= option to 'YES', which prevents the titles and footnotes from wrapping.

SAS Code 8-18 – ODS Excel TITLE_FOOTNOTE_NOBREAK= Option Set to YES

```
title    "This a my long title so I can show that it does not wrap
around when title_footnote_nobreak=''yes''";


footnote "This a my very long footnote (it prints in a smaller font)
so I can show that it does not wrap around when
title_footnote_nobreak=''yes''";
ods excel file = "&path\TITLE_FOOTNOTE_NOBREAK_yes.xlsx"
          options(title_footnote_nobreak='yes'
                  embedded_titles='yes'
                  embedded_footnotes='yes');
 proc print data=asia_only;
  run;
ods excel close;
```

The SAS Code 8-18 code is set to YES for the TITLE_FOOTNOTE_NOBREAK= option and causes the titles and footnotes not to wrap to the next line if the title or footnote is too long.

Figure 8-35 – The TITLE_FOOTNOTE_NOBREAK= Option Set to "YES"

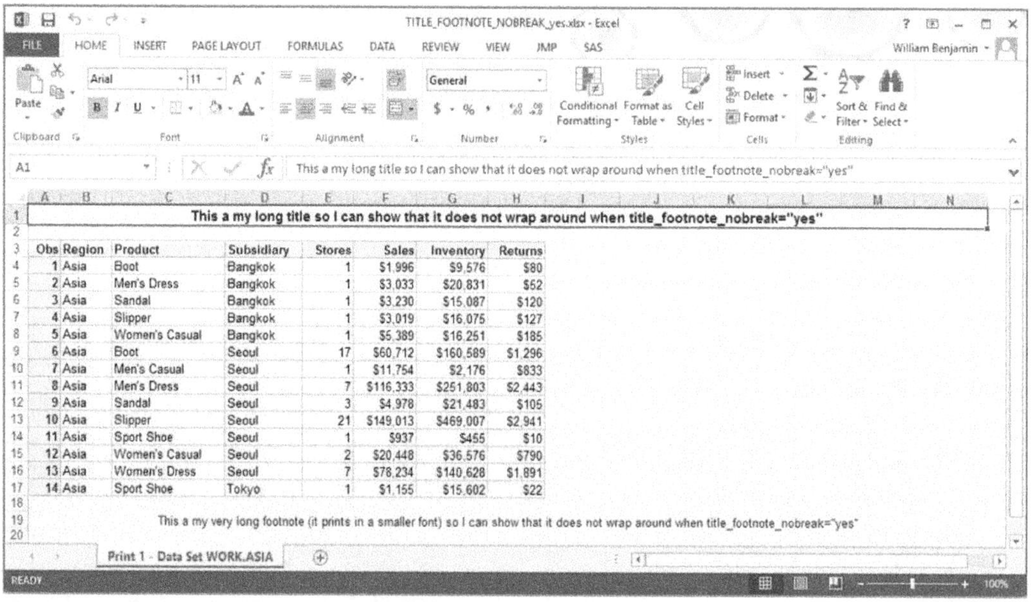

Obs	Region	Product	Subsidiary	Stores	Sales	Inventory	Returns
1	Asia	Boot	Bangkok	1	$1,996	$9,576	$80
2	Asia	Men's Dress	Bangkok	1	$3,033	$20,831	$52
3	Asia	Sandal	Bangkok	1	$3,230	$15,087	$120
4	Asia	Slipper	Bangkok	1	$3,019	$16,075	$127
5	Asia	Women's Casual	Bangkok	1	$5,389	$16,251	$185
6	Asia	Boot	Seoul	17	$60,712	$160,589	$1,296
7	Asia	Men's Casual	Seoul	1	$11,754	$2,176	$833
8	Asia	Men's Dress	Seoul	7	$116,333	$251,803	$2,443
9	Asia	Sandal	Seoul	3	$4,978	$21,483	$105
10	Asia	Slipper	Seoul	21	$149,013	$469,007	$2,941
11	Asia	Sport Shoe	Seoul	1	$937	$455	$10
12	Asia	Women's Casual	Seoul	2	$20,448	$36,576	$790
13	Asia	Women's Dress	Seoul	7	$78,234	$140,628	$1,891
14	Asia	Sport Shoe	Tokyo	1	$1,155	$15,602	$22

When using the ODS EXCEL TITLE_FOOTNOTE_NOBREAK= option set to NO, the titles and footnotes do not wrap to the next line if the title or footnote is too long. Also, note that the titles and footnotes are printed in different font sizes and have different bold options.

The TITLE_FOOTNOTE_WIDTH= Option

The TITLE_FOOTNOTE_WIDTH= option describes the number of columns that a SAS title or footnote can span. The default value of zero means that the titles and footnotes can span the same number of columns as the column data that is output. Here I will use long titles and footnotes to show the effects. Note that if the number of columns is too small the output might appear to be cutoff because the display space of the Excel cells are too small.

SAS Code 8-19 – ODS Excel TITLE_FOOTNOTE_WIDTH= Option Set to the Default

```
title    "This a my long title so I can show that it wraps around
when title_footnote_width is not active.";
footnote "This a my very long footnote (it prints in a smaller font)
so I can show that it wraps around when title_footnote_width not
active.";

ods excel file = "&path\TITLE_FOOTNOTE_WIDTH_default.xlsx"
         options(embedded_titles='yes'
                embedded_footnotes='yes');
 proc print data=asia_only;
  run;
ods excel close;
```

Not using the TITLE_FOOTNOTE_WIDTH= option is the same as setting TITLE_FOOTNOTE_WIDTH= 0. This sets the title and footnote width to the number of data columns that are output.

Figure 8-36 – The TITLE_FOOTNOTE_WIDTH= Option Set to the Default by Not Using It

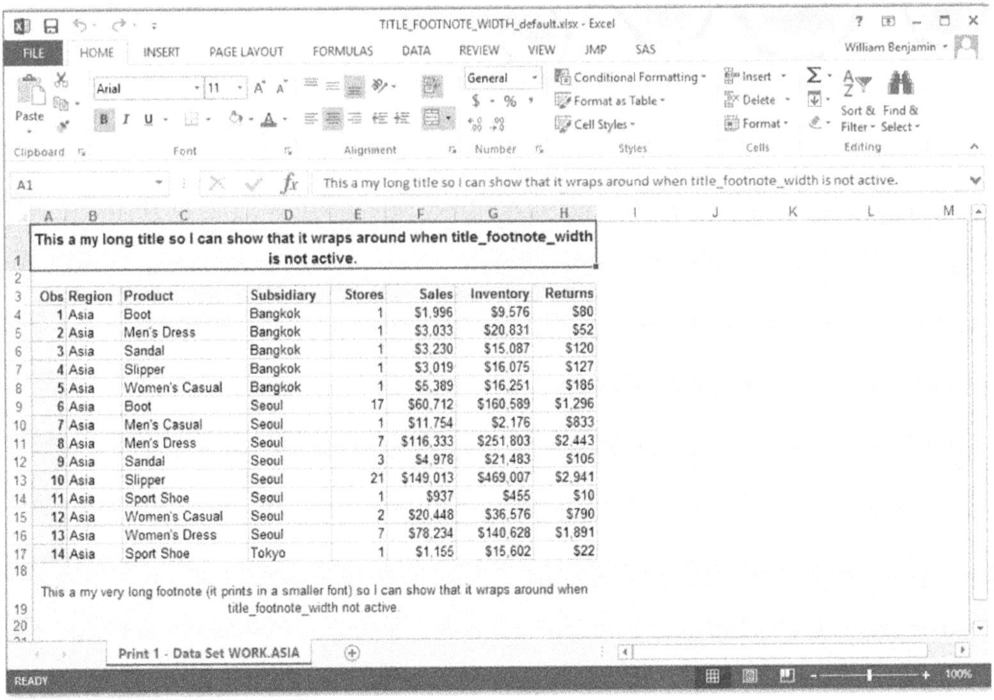

Obs	Region	Product	Subsidiary	Stores	Sales	Inventory	Returns
1	Asia	Boot	Bangkok	1	$1,996	$9,576	$80
2	Asia	Men's Dress	Bangkok	1	$3,033	$20,831	$52
3	Asia	Sandal	Bangkok	1	$3,230	$15,087	$120
4	Asia	Slipper	Bangkok	1	$3,019	$16,075	$127
5	Asia	Women's Casual	Bangkok	1	$5,389	$16,251	$185
6	Asia	Boot	Seoul	17	$60,712	$160,589	$1,296
7	Asia	Men's Casual	Seoul	1	$11,754	$2,176	$833
8	Asia	Men's Dress	Seoul	7	$116,333	$251,803	$2,443
9	Asia	Sandal	Seoul	3	$4,978	$21,483	$105
10	Asia	Slipper	Seoul	21	$149,013	$469,007	$2,941
11	Asia	Sport Shoe	Seoul	1	$937	$455	$10
12	Asia	Women's Casual	Seoul	2	$20,448	$36,576	$790
13	Asia	Women's Dress	Seoul	7	$78,234	$140,628	$1,891
14	Asia	Sport Shoe	Tokyo	1	$1,155	$15,602	$22

This a my very long footnote (it prints in a smaller font) so I can show that it wraps around when title_footnote_width not active.

By not using the TITLE_FOOTNOTE_WIDTH= option or setting it to zero, the width is set to the number of columns of data output.

SAS Code 8-20 – ODS Excel TITLE_FOOTNOTE_WIDTH= Option Set to 4

```
title    "This a my long title so I can show how it wraps around
when title_footnote_width=''4''";
footnote "This a my very long footnote (it prints in a smaller font)
so I can hwshow that it wraps around when
title_footnote_width=''4''";
ods excel file = "&path\TITLE_FOOTNOTE_WIDTH_4.xlsx"
            options(title_footnote_width='4'
                    embedded_titles='yes'
                    embedded_footnotes='yes');
 proc print data=asia_only;
   run;
ods excel close;
```

Note that providing a title or a footnote that is too long can cause the title or footnote to wrap too many times, then not all of the title or footnote will be visible in the space provided. You will usually need to correct this by increasing the size of the cell in Excel.

Figure 8-37 – The TITLE_FOOTNOTE_WIDTH= Option Set to 4

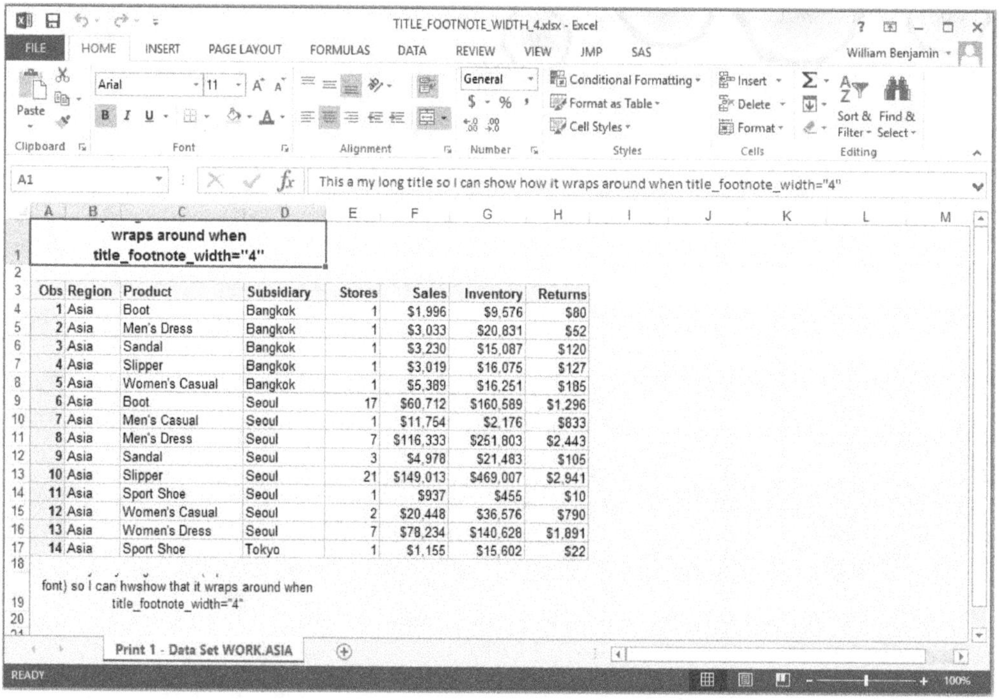

Here the TITLE_FOOTNOTE_WIDTH= value is set to "4" which causes the long titles and footnotes to wrap within the first four columns. This has the effect of hiding some of the text. But all of the text is available within the assigned title and footnote cells.

SAS Code 8-21 – ODS Excel TITLE_FOOTNOTE_WIDTH= Option Set to 10

```
title    "This a my long title so I can show that it does not wrap
around when title_footnote_width=''10''";
footnote "This a my very long footnote (it prints in a smaller font)
so I can show that it does not wrap around when
title_footnote_width=''10''";
ods excel file = "&path\TITLE_FOOTNOTE_WIDTH_10.xlsx"
            options(title_footnote_width='10'
                      embedded_titles='yes'
                      embedded_footnotes='yes');
 proc print data=asia_only;
  run;
ods excel close;
```

This example applies a title/footnote width that is larger than the number of data columns printed, but if either the Title or Footnote is too long, it will wrap to the next line.

Figure 8-38 – The TITLE_FOOTNOTE_WIDTH= Option Set to 10

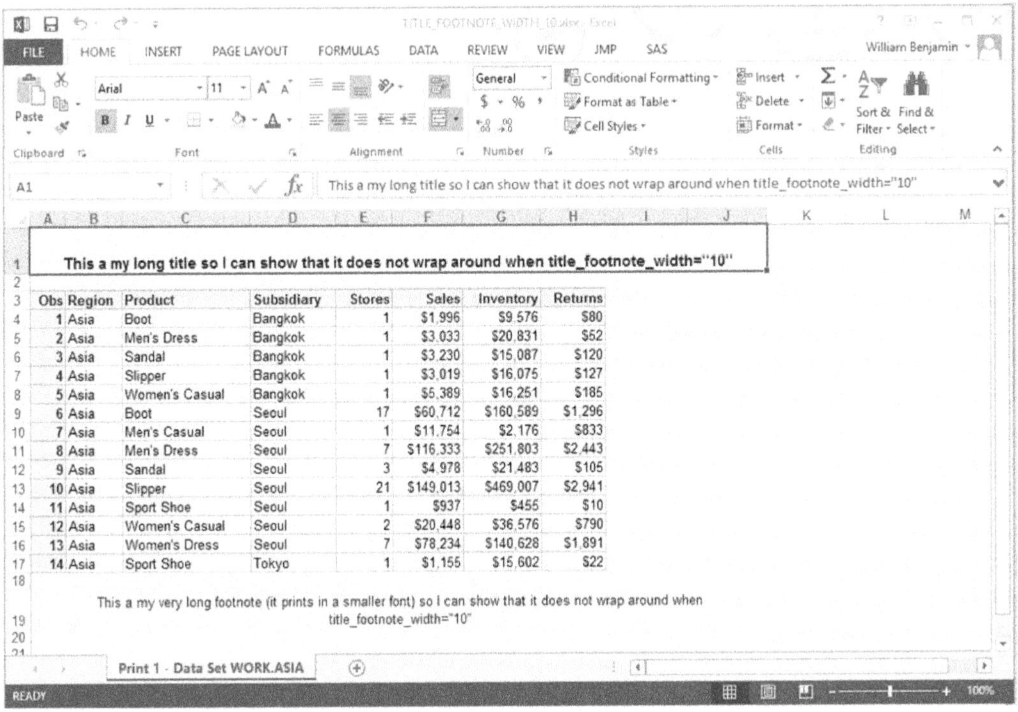

Obs	Region	Product	Subsidiary	Stores	Sales	Inventory	Returns
1	Asia	Boot	Bangkok	1	$1,996	$9,576	$80
2	Asia	Men's Dress	Bangkok	1	$3,033	$20,831	$52
3	Asia	Sandal	Bangkok	1	$3,230	$15,087	$120
4	Asia	Slipper	Bangkok	1	$3,019	$16,075	$127
5	Asia	Women's Casual	Bangkok	1	$5,389	$16,251	$185
6	Asia	Boot	Seoul	17	$60,712	$160,589	$1,296
7	Asia	Men's Casual	Seoul	1	$11,754	$2,176	$833
8	Asia	Men's Dress	Seoul	7	$116,333	$251,803	$2,443
9	Asia	Sandal	Seoul	3	$4,978	$21,483	$105
10	Asia	Slipper	Seoul	21	$149,013	$469,007	$2,941
11	Asia	Sport Shoe	Seoul	1	$937	$455	$10
12	Asia	Women's Casual	Seoul	2	$20,448	$36,576	$790
13	Asia	Women's Dress	Seoul	7	$78,234	$140,628	$1,891
14	Asia	Sport Shoe	Tokyo	1	$1,155	$15,602	$22

Here, with a value of 10, the titles and footnotes extend beyond the data columns. As you can see they will still wrap if the length of the title or footnote is too long.

Conclusion

This chapter has shown a lot of different ways to manipulate the output features of Excel from within your SAS code. While working with only the features to change or shift the output data, you can perform many tasks that you have been doing by hand, after the workbooks and spreadsheets are created. If you are doing this for a one-time report for a weekly meeting to give to your boss, all of this effort to find and use these ODS Excel destination suboptions might be over-kill. But, if you are producing dozens or hundreds of reports, these suboptions can be considered useful.

Chapter 9: Column, Row, and Cell Features

Introduction

In this chapter, we will discuss the ODS EXCEL options that affect Row, Column, and Cell level data that is output into an Excel worksheet. In this chapter, the SASHELP.SHOES data set will be filtered with a WHERE clause to include only data from the "ASIA" region along with part of the SASHELP.ORSALES data set. This will enable me to show all of the data on the same worksheet. Unless otherwise noted, all of the examples in this chapter will be using the following data set called "ASIA_ONLY" that was shown in Chapter 2.

ODS Excel Destination Column, Row, and Cell Options

In this chapter, we will discuss the following SAS ODS EXCEL destination topics. The Microsoft Help screens provide a description of how the column width and row heights are determined, which I have included here to help describe the heights listed in the following tables.

"On a worksheet, you can specify a column width of 0 (zero) to 255. This value represents the number of characters that can be displayed in a cell that is formatted with the standard font. The default column width is 8.43 characters. If a column has a width of 0 (zero), the column is hidden.

You can specify a row height of 0 (zero) to 409. This value represents the height measurement in points (1 point equals approximately 1/72 inch or 0.035 cm). The default row height is 12.75 points (approximately 1/6 inch or 0.4 cm). If a row has a height of 0 (zero), the row is hidden."

Table 9.1 – ODS Excel Destination Suboptions for Column Features

Column Features		
Action Parameter	**Options**	**Description**
ABSOLUTE_COLUMN_WIDTH	'number-list' or 'NONE' Default = 'NONE'	This option enables you to specify the width of columns of the output spreadsheet by providing a quoted string of numbers separated by commas, or the quoted string 'NONE'. Multiple values produce a repeating pattern of the specified column widths.
AUTOFILTER	'ALL', 'NONE', or 'range' Default = 'NONE'	This option enables you to apply an Excel filter option to all columns of an Excel worksheet, a range of columns, or to none of the columns of the output worksheet.
COLUMN_REPEAT	'number', 'number-range', or 'HEADER' Default = no header repeating	This option affects the printing of column headings across pages. The 'HEADER' option causes headers to be printed for all columns that contain header values on each page. While the 'number' and 'number-range' options turn on printing for either a column or range of columns, respectively.
FROZEN_HEADERS	'ON', 'OFF', 'TRUE', 'FALSE', 'YES','NO', number Default = 'OFF'	Positive responses freeze the header row so that when the scroll bar is moved the headers remain visible. Providing a number freezes the specified number of rows.
HIDDEN_COLUMNS	'number_list_range', Default = 'NONE'	Identifies what columns are to be hidden. The number_range_list is a quoted string of numbers or number ranges that are separated by commas.

Table 9.2 – ODS Excel Destination Suboptions for Row Features

Action Parameter	Row Features Options	Description
ABSOLUTE_ROW_HEIGHT	'number_list', 'number-range', 'OFF', 'FALSE', Default = 'NONE'	This option allows the user to override the row heights generated by SAS by providing a quoted string of values separated by commas. Multiple heights can be provided. Multiple values produce a repeating pattern of the specified row heights.
FROZEN_ROWHEADERS	'ON', 'OFF', 'TRUE', 'FALSE', 'YES','NO', number Default = 'OFF'	Positive responses freeze the header column (Column one) so that when the scroll bar is moved the row headers remain visible. Providing a number freezes the specified number of columns.
HIDDEN_ROWS	'number_list_range', Default = Show all rows	Identifies what rows are to be hidden. The number_range_list is a quoted string of numbers or number ranges that are separated by commas.
ROW_HEIGHTS	'number_list', Default = use font size to define row heights	Provides a quoted string of positional parameters to identify row heights (0 = use the font height) for the following (order-values might be skipped but skipped parameters require a comma to show the parameter is missing. Except the last used value.) • Table header rows • Table body rows • BY value lines • Titles • Footers • Page break height • Paragraph skip height
ROW_REPEAT	'number-range', Default = 'NONE'	This option specifies the way that row headers are processed across pages. • 'NONE' – No rows are repeated. • 'HEADER' – Repeat all row headings. • 'number' – Repeat row

Action Parameter	Row Features	
	Options	Description
		header for this row.
		• 'number-range' – Repeat rows within the specified range.

Table 9.3 – ODS Excel Destination Suboptions for Cell Features

Action Parameter	Cell-Level Features	
	Options	Description
FORMULAS	'ON', 'OFF', 'TRUE', 'FALSE', 'YES','NO', number Default = 'ON'	This option enables you to change the SAS output to the default Excel formula feature. Cells that begin with an equal sign (=) can be processed as a formula by Excel. Positive responses allow all the formula processing. Negative responses turn off the SAS processing that creates formulas when the cell values begin with an equal sign.
START_AT	'x,y' where x is a row number and y is a column number. Default = '1,1'	A string that indicates the starting ROW and COLUMN in the Excel worksheet where the output data is to start being output. This cannot be changed within a sheet. SAS 9.4M4 allows the use of alphabetic characters for the 'Column' value as in 'D4'.

The ABSOLUTE_COLUMN_WIDTH= Suboption

Often times you will find that you have data that either you or your boss want to see in one cell, for example a product description. But if a count of inventory or bin number (in the same column or in a different one?) is only a few characters long, you might also want to limit the size of that column. Most Excel output files have variable widths for the data in columns. While Excel does a fair job of adjusting the column widths when data is read into the workbook sheets, it often truncates or pads the width based on either the first row (variable titles or column names), or some other formula known only to Excel.

The ABSOLUTE_COLUMN_WIDTH= option enables you to adjust the widths of the columns based on your needs. This enables you to take control of one more part of the output to make it what you want. Figure 9-1 shows the output of the ODS Excel destination without modifications for the SAS WORK.ASIA_ONLY data set. Figure 9.2 shows the output with adjusted column widths. SAS Code 9.1 shows the code to produce output from WORK.ASIA_ONLY with and without ABSOLUTE_COLUMN_WIDTH adjustments.

SAS Code 9-1 – ODS Excel Code to Show Default ABSOLUTE_COLUMN_WIDTH= Usage

```
ods excel file = "&path\Non_Adjusted_column_width.xlsx"
  options(ABSOLUTE_COLUMN_WIDTH='NONE');
  proc print data=ASIA_ONLY noobs;
  run;
ods excel close;
```

Figure 9-1 – Excel Output without Adjusted Column Widths

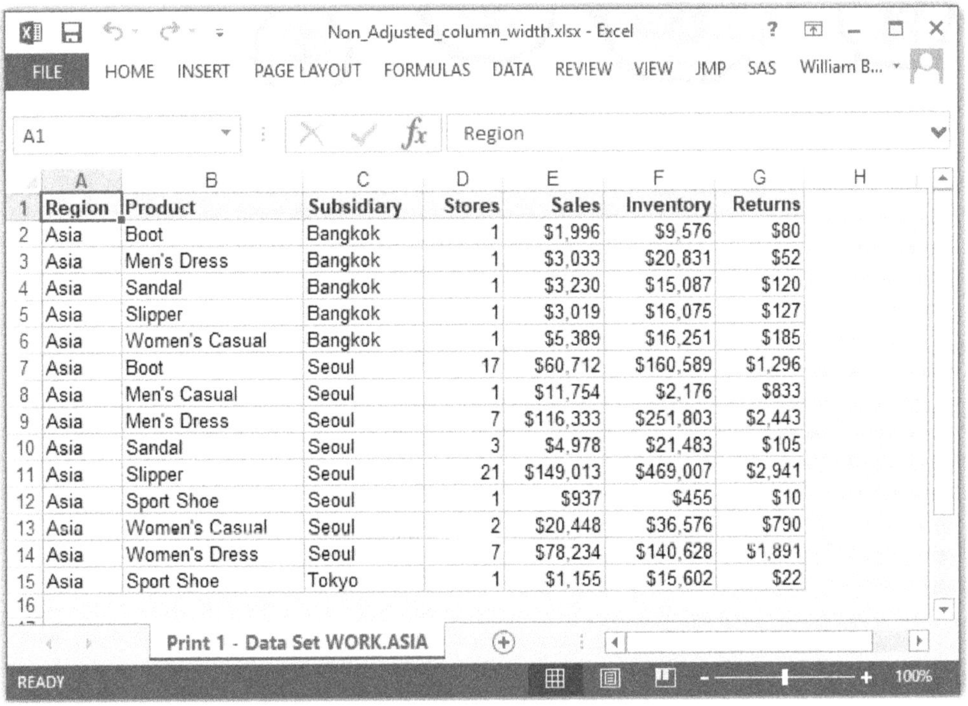

In Figure 9-1 many of the column titles have spaces in front of the column names in row one of the Excel worksheet.

SAS Code 9-2– ODS Excel Code to Show ABSOLUTE_COLUMN_WIDTH= Usage

```
ods excel file = "&path\Adjusted_column_width.xlsx"
        options(ABSOLUTE_COLUMN_WIDTH='6,14,10,6,8,9,8');
  proc print data= ASIA_ONLY noobs;
  run;
ods excel close;
```

As mentioned above, the numbers in the ABSOLUTE_COLUMN_WIDTH= suboption in SAS Code 9-2 are column-specific character counts that reduce the column widths.

Figure 9-2 – Excel Output with Adjusted Column Widths

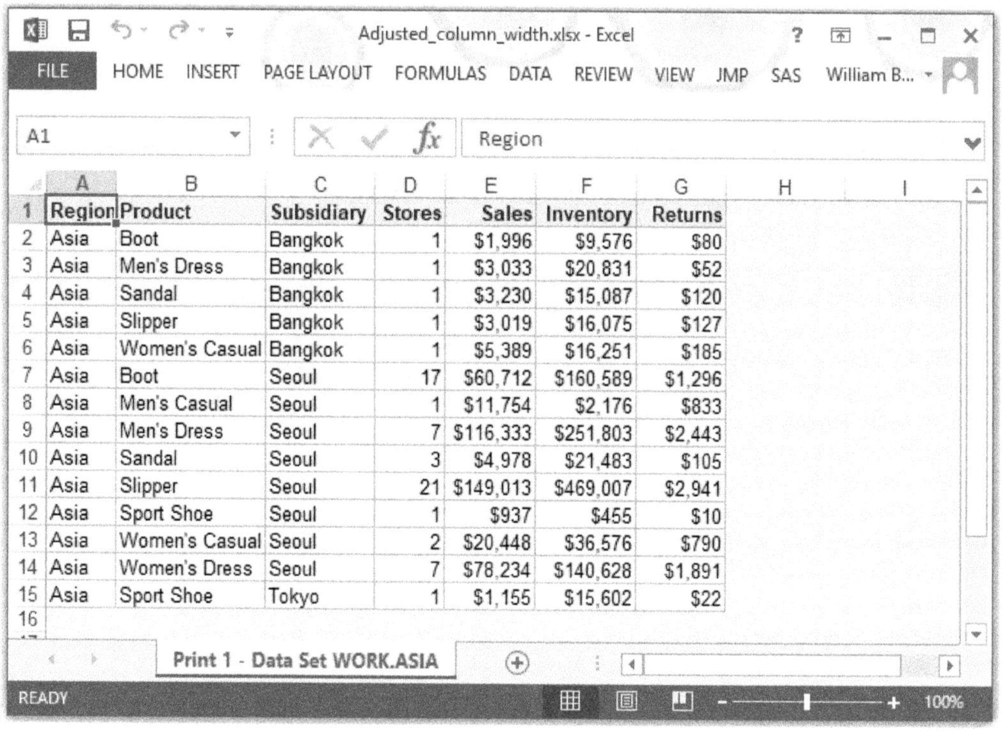

	A	B	C	D	E	F	G	H	I
1	Region	Product	Subsidiary	Stores	Sales	Inventory	Returns		
2	Asia	Boot	Bangkok	1	$1,996	$9,576	$80		
3	Asia	Men's Dress	Bangkok	1	$3,033	$20,831	$52		
4	Asia	Sandal	Bangkok	1	$3,230	$15,087	$120		
5	Asia	Slipper	Bangkok	1	$3,019	$16,075	$127		
6	Asia	Women's Casual	Bangkok	1	$5,389	$16,251	$185		
7	Asia	Boot	Seoul	17	$60,712	$160,589	$1,296		
8	Asia	Men's Casual	Seoul	1	$11,754	$2,176	$833		
9	Asia	Men's Dress	Seoul	7	$116,333	$251,803	$2,443		
10	Asia	Sandal	Seoul	3	$4,978	$21,483	$105		
11	Asia	Slipper	Seoul	21	$149,013	$469,007	$2,941		
12	Asia	Sport Shoe	Seoul	1	$937	$455	$10		
13	Asia	Women's Casual	Seoul	2	$20,448	$36,576	$790		
14	Asia	Women's Dress	Seoul	7	$78,234	$140,628	$1,891		
15	Asia	Sport Shoe	Tokyo	1	$1,155	$15,602	$22		
16									

Print 1 - Data Set WORK.ASIA

When the true sizes of the output fields are known, then you can adjust the width of the Excel columns to use the space on the spreadsheet more efficiently.

The AUTOFILTER= Suboption

The Excel AUTOFILTER= option is something that is useful within the Excel workbook. It enables you to filter and sort values within a column of data without modifying the Excel worksheet. The default is to have all of the columns not filtered. An alternative strategy is to have all of the columns filtered. This option goes one step further in that it enables you to select a range of columns to filter. You can choose one column or a set of contiguous columns to filter. My examples in SAS Code 9-3 and SAS Code 9-4 show the 'all' option and the range option but do not show the null condition or a single column. The single column option would be a number range of just one number.

SAS Code 9-3 – Code to Apply to Filters to All Excel Columns

```
ods excel file = "&path\AUTOFILTER_all.xlsx"
          options(AUTOFILTER='all');
  proc print data=Asia_only;
  run;
ods excel close;
```

Figure 9-3 – ODS AUTOFILTER= Suboption for All Columns

The small down arrow on the first row of the spreadsheet indicates that the column is filtered.

SAS Code 9-4 – Code to Apply to Filters to a Range of Excel Columns

```
ods excel file = "&path\AUTOFILTER_range.xlsx"
       options(AUTOFILTER='3-5');
  proc print data=Asia_only;
  run;
ods excel close;
```

Figure 9-4 – AUTOFILTER= to Filter a Range of Columns

	Obs	Region	Product	Subsidiary	Stor	Sales	Inventory	Returns
1	Obs	Region	Product	Subsidiary	Stor	Sales	Inventory	Returns
2	1	Asia	Boot	Bangkok	1	$1,996	$9,576	$80
3	2	Asia	Men's Dress	Bangkok	1	$3,033	$20,831	$52
4	3	Asia	Sandal	Bangkok	1	$3,230	$15,087	$120
5	4	Asia	Slipper	Bangkok	1	$3,019	$16,075	$127
6	5	Asia	Women's Casual	Bangkok	1	$5,389	$16,251	$185
7	6	Asia	Boot	Seoul	17	$60,712	$160,589	$1,296
8	7	Asia	Men's Casual	Seoul	1	$11,754	$2,176	$833
9	8	Asia	Men's Dress	Seoul	7	$116,333	$251,803	$2,443
10	9	Asia	Sandal	Seoul	3	$4,978	$21,483	$105
11	10	Asia	Slipper	Seoul	21	$149,013	$469,007	$2,941
12	11	Asia	Sport Shoe	Seoul	1	$937	$455	$10
13	12	Asia	Women's Casual	Seoul	2	$20,448	$36,576	$790
14	13	Asia	Women's Dress	Seoul	7	$78,234	$140,628	$1,891
15	14	Asia	Sport Shoe	Tokyo	1	$1,155	$15,602	$22

Notice here that only three columns, 'C', 'D', and 'E' have filtering turned on.

The COLUMN_REPEAT= Suboption

The ODS Excel destination COLUMN_REPEAT= option is a little tricky to identify. This option enables you to copy some information onto the subsequent sheets of a wide printout. If your output file that you send to Excel does not have enough columns to cause a printout wide enough to extend to a second page, then the option does not appear to do anything. Perhaps you have a lot of information about a part or inventory list and you want to have the product description on the second half or following pages across the printout. In this example I will be using the first 53 rows of the SASHELP.ORSALES SAS data set. The Excel "Print Preview" output displayed below will be shown as the left and right halves of the data. You can use 'NONE,' a number, a number-range, or 'HEADER' to describe what you want to repeat. Also, using the number '1' is the same as using 'HEADER'. Neither FROZEN_HEADERS= nor FROZEN_ROWHEADERS= affect the number of header columns.

SAS Code 9-5 – Use of the COLUMN_REPEAT= Suboption for One Column

```
ods excel file = "&path\COLUMN_REPEAT_number.xlsx"
          options(COLUMN_REPEAT='6');
  proc print data=SASHELP.ORSALES(obs=53);
  run;
ods excel close;
```

Figure 9-5 – Excel Page Setup Sheet Showing Column "$F" Is Repeated

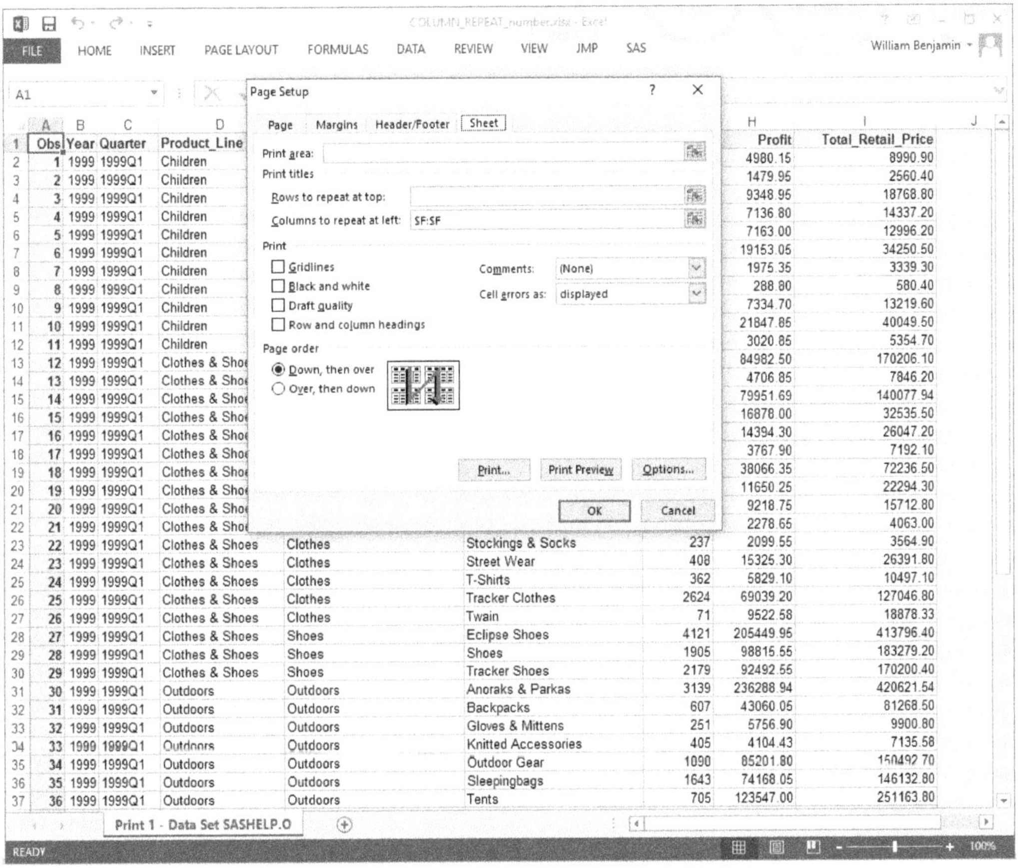

The notation in the "Columns to repeat at left:" shows "$F:$F"; the variable PRODUCT_LINE will be repeated when the second half of the page is printed. The "$" sign indicates the absolute row "F". The "Page order" selection specifies that all data from the columns that fit onto the first page will be printed before any data will be shown on the second page. I carefully chose 53 data lines in this example so that all COLUMN_REPEAT= option output filled exactly one page. The data on page two is the second half of the output.

Figure 9-6 –First Preview Page of the COLUMN_REPEAT Column '6' Output

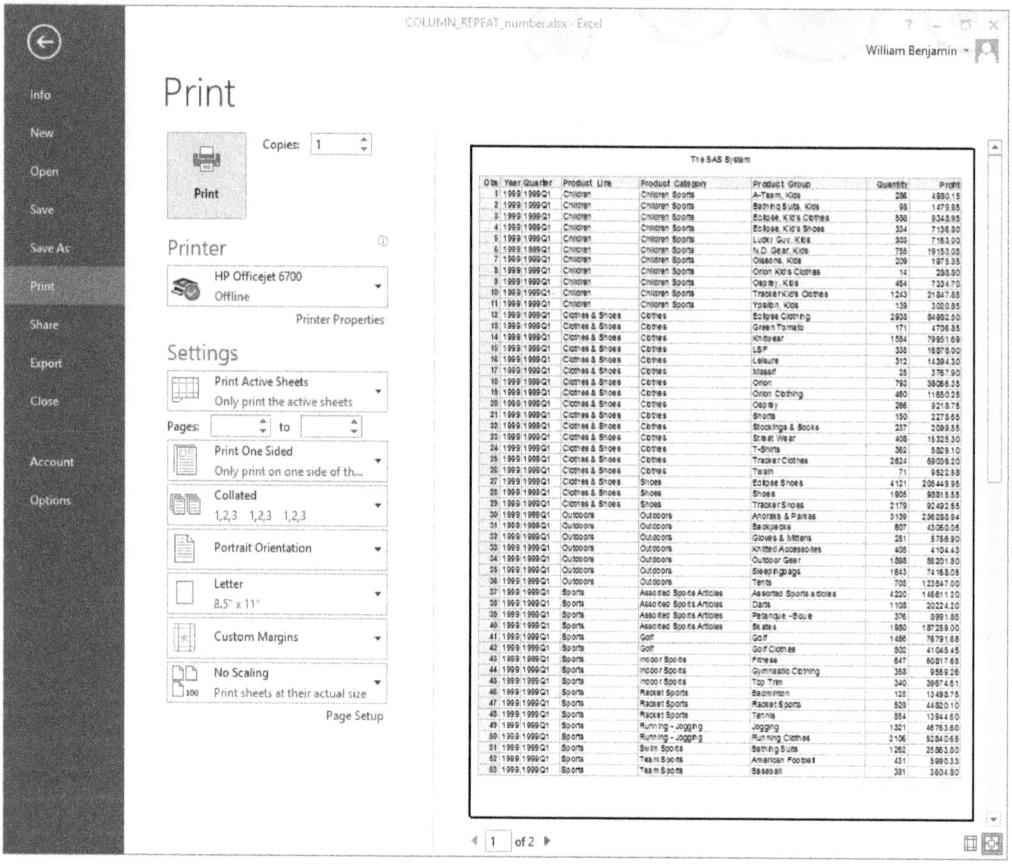

This page shows the first page of the data as it would print. Column "6" is the "Product_Group" variable.

Figure 9-7 – Second Preview Page of the COLUMN_REPEAT Column '6' Output

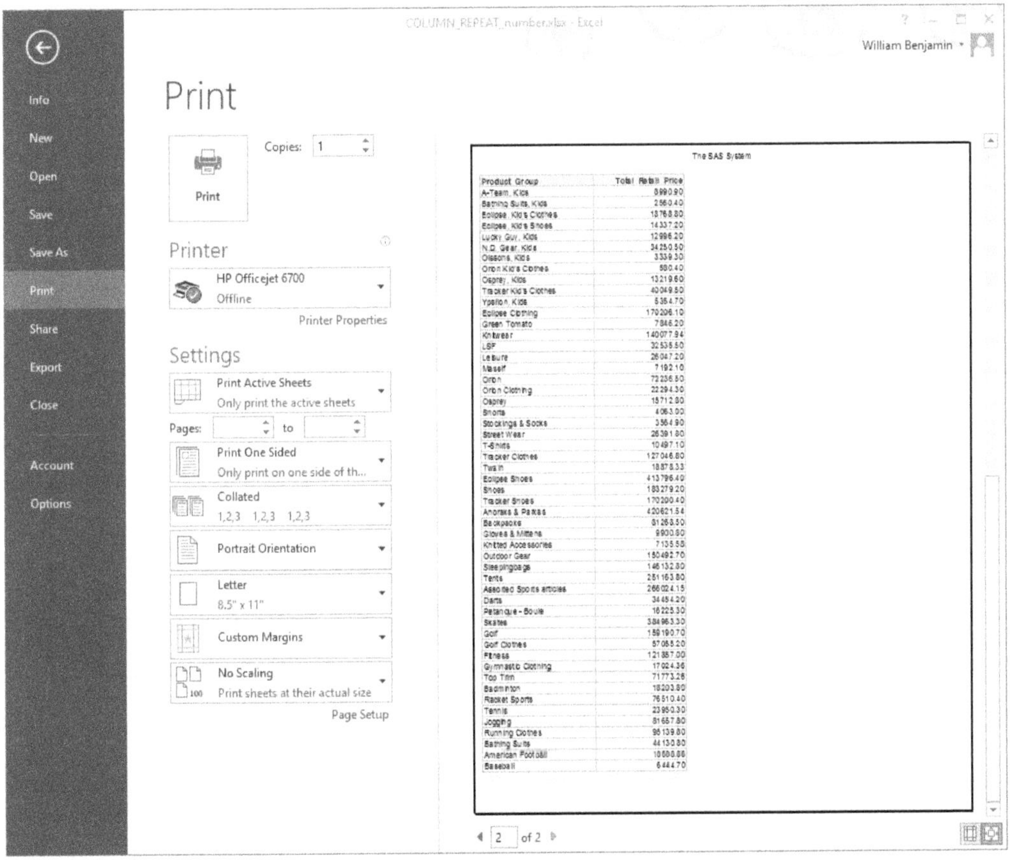

This page shows the second page of the data as it would print. Column "6" ("Product_Group") is repeated as the first column on this page.

SAS Code 9-6 – Use of the COLUMN_REPEAT= Suboption for a Column Range

```
ods excel (id=2) file = "&path\COLUMN_REPEAT_range.xlsx"
          options(COLUMN_REPEAT='3-5');
  proc print data=SASHELP.ORSALES(obs=53);
  run;
ods excel close;
```

Figure 9-8 –COLUMN_REPEAT= Excel "Page Setup" Display for a Range of Columns

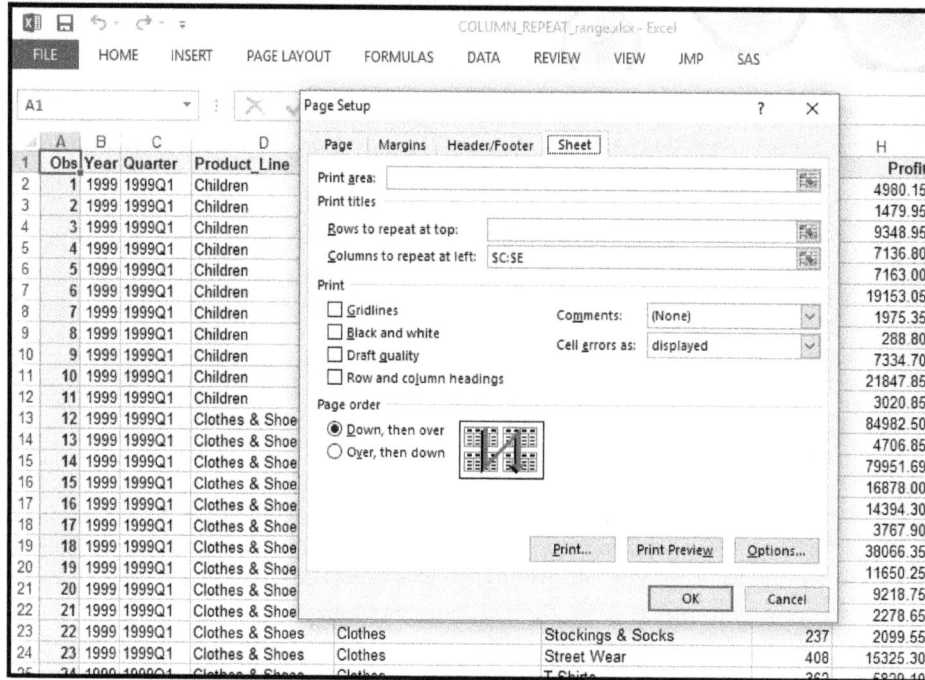

Here columns "C" through "E" are to be printed on the second half of the output.

Figure 9-9 – Partial Image of the Second Half of the Output

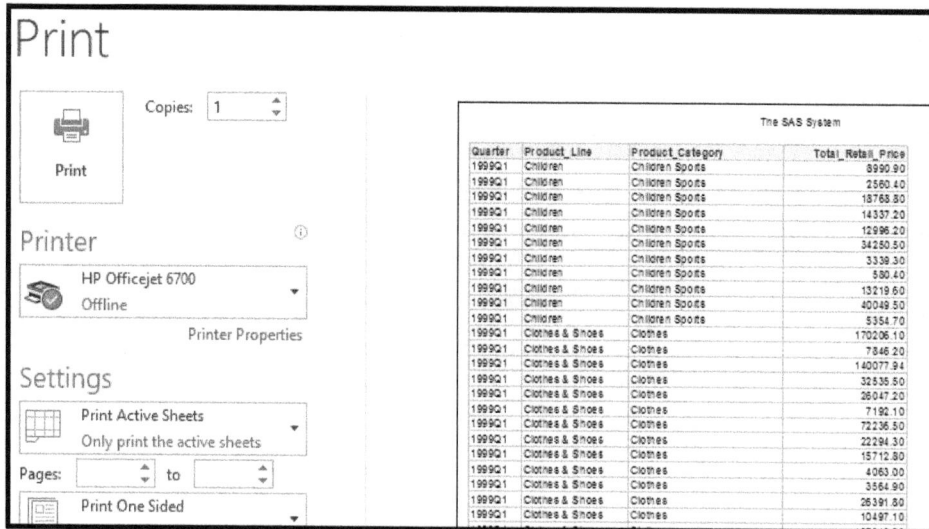

The column "Quarter", "Product_Line", and "Product_Category" are repeated on the second half of the output.

SAS Code 9-7 – Use of the COLUMN_REPEAT= Suboption HEADER Value

```
ods excel file = "&path\COLUMN_REPEAT_header.xlsx"
        options(COLUMN_REPEAT='header');
  proc print data=SASHELP.ORSALES(obs=53);
  run;
ods excel close;
```

Figure 9-10 – Sample Data When the COLUMN_REPEAT= "HEADER" Option Is Used

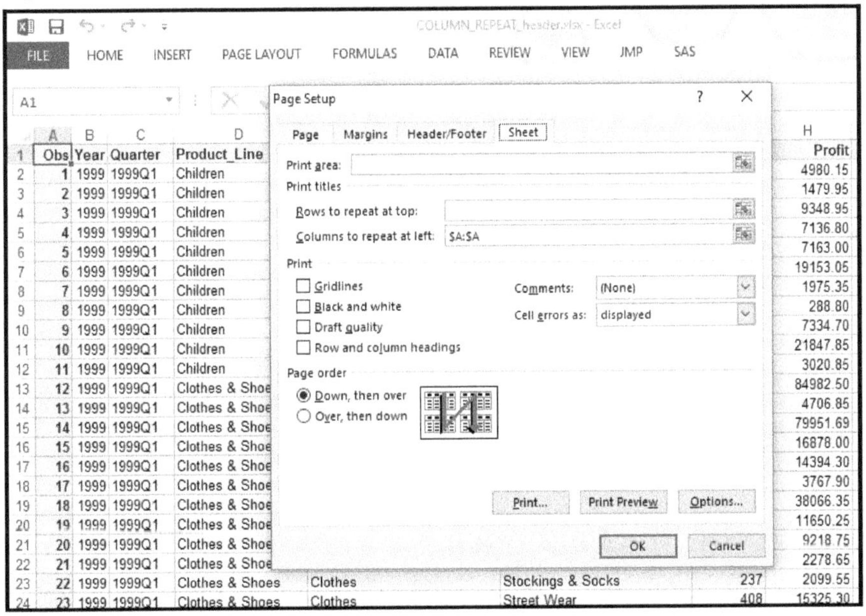

Column "A" is always considered the "HEADER" and neither the FROZEN_HEADERS= nor FROZEN_ROWHEADERS= affect the number of header columns.

Figure 9-11 – COLUMN_REPEAT Partial Image of Header Output

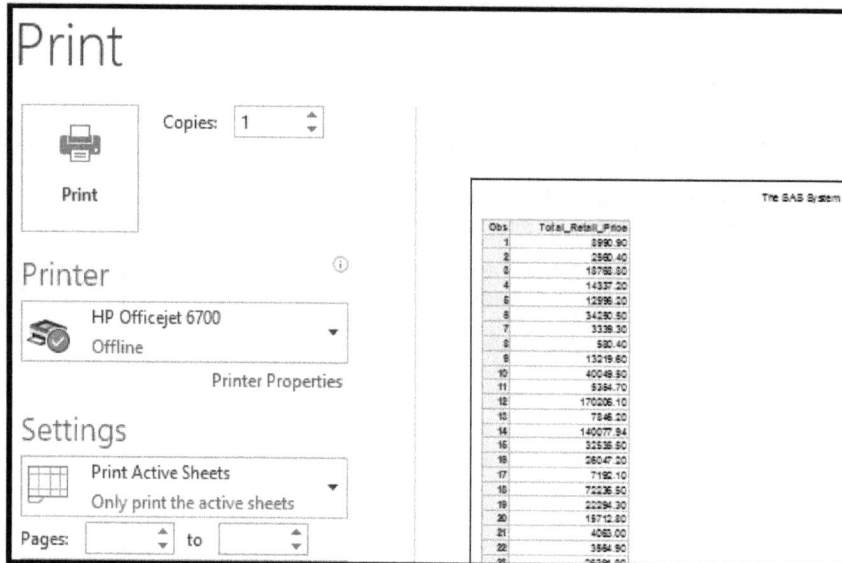

Here the 'HEADER' option reproduces column "A" of the spreadsheet onto the second half of the printed output.

The FROZEN_HEADERS= Suboption

The ODS Excel FROZEN_HEADERS= option does another thing that I used to do by hand for years. It sets up the top rows of the Excel worksheet so that they do not scroll. The choices are either all rows scrolling (the default) or one or more rows frozen at the top of the worksheet. The values "ON" or the number "1" freeze only the top row, any other number freezes that many rows.

SAS Code 9-8 – The FROZEN_HEADERS= Suboption Default Value

```
ods excel file = "&path\Frozen_Headers_on.xlsx"
          options(Frozen_Headers='on');
  proc print data=Asia_only;
  run;
ods excel close;
```

Figure 9-12 – The FROZEN_HEADERS= Suboption "ON" Output

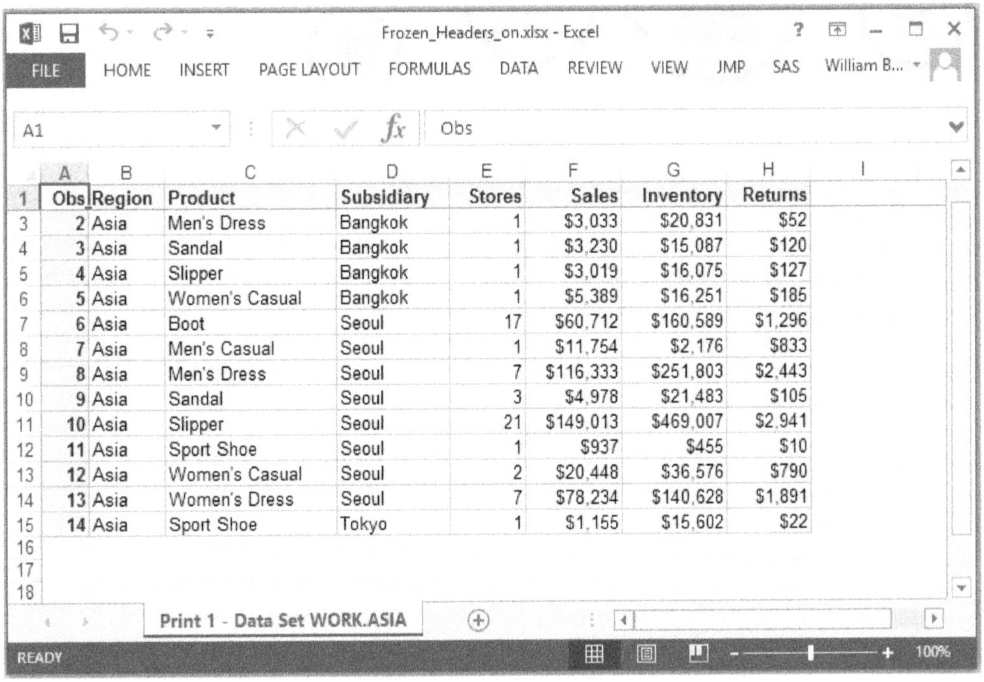

	A	B	C	D	E	F	G	H	I
1	Obs	Region	Product	Subsidiary	Stores	Sales	Inventory	Returns	
3	2	Asia	Men's Dress	Bangkok	1	$3,033	$20,831	$52	
4	3	Asia	Sandal	Bangkok	1	$3,230	$15,087	$120	
5	4	Asia	Slipper	Bangkok	1	$3,019	$16,075	$127	
6	5	Asia	Women's Casual	Bangkok	1	$5,389	$16,251	$185	
7	6	Asia	Boot	Seoul	17	$60,712	$160,589	$1,296	
8	7	Asia	Men's Casual	Seoul	1	$11,754	$2,176	$833	
9	8	Asia	Men's Dress	Seoul	7	$116,333	$251,803	$2,443	
10	9	Asia	Sandal	Seoul	3	$4,978	$21,483	$105	
11	10	Asia	Slipper	Seoul	21	$149,013	$469,007	$2,941	
12	11	Asia	Sport Shoe	Seoul	1	$937	$455	$10	
13	12	Asia	Women's Casual	Seoul	2	$20,448	$36,576	$790	
14	13	Asia	Women's Dress	Seoul	7	$78,234	$140,628	$1,891	
15	14	Asia	Sport Shoe	Tokyo	1	$1,155	$15,602	$22	

Here I have scrolled the Excel worksheet data up from the bottom by one line. This is noticed by the fact that row '2' is not shown. The default output from SAS does not show grid lines outside of the data output by SAS.

SAS Code 9-9 – The FROZEN_HEADERS= Suboption Freezing Four Rows

```
ods excel file = "&path\Frozen_Headers_number.xlsx"
          options(Frozen_Headers='4');
  proc print data=Asia_only;
  run;
ods excel close;
```

The code in SAS Code 9-9 freezes the top four rows of the Excel output worksheet. The image in Figure 9-13 is the output of the executed SAS code.

Figure 9-13 – The FROZEN_HEADERS= Suboption as a Number Set to "4"

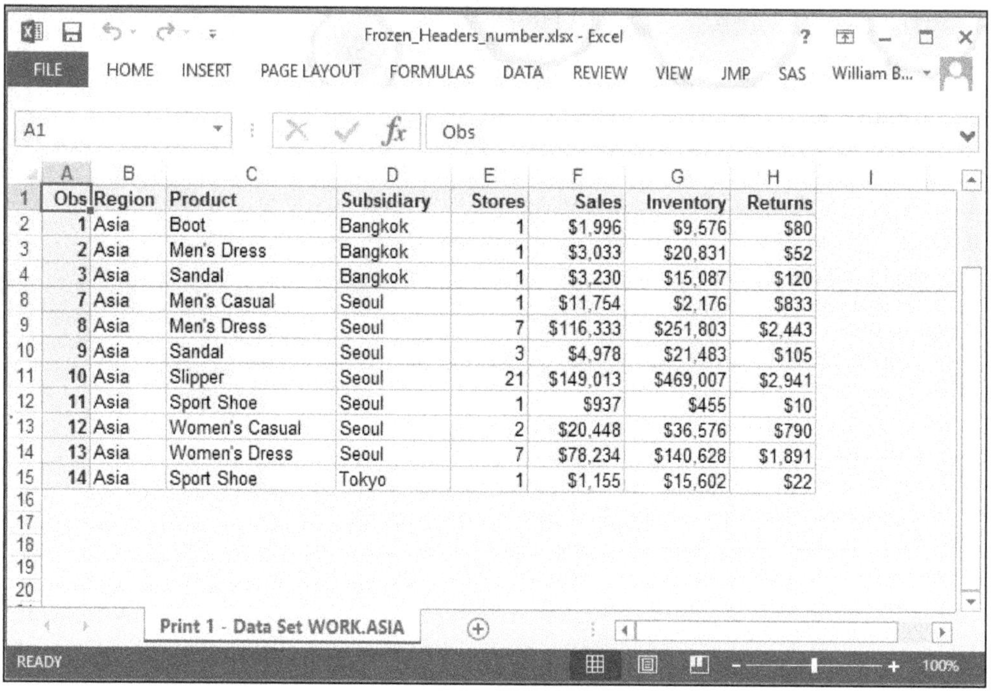

Once again there are missing rows in the display indicated by the visible horizontal line after row "4". I scrolled the data up three rows to make this image.

The HIDDEN_COLUMNS= Suboption

Hiding columns in an Excel worksheet is another task often done manually after the workbooks are created. Again, SAS can do this for you when the worksheets are created. The HIDDEN_COLUMNS= option is very flexible, because you can combine individual columns and ranges in the same 'number_list_range' value string. The default option is to not have any columns hidden.

SAS Code 9-10 – Using the HIDDEN_COLUMNS= Option

```
ods excel file = "&path\Hidden_Columns.xlsx"
        options(Hidden_Columns='2,5-6');
  proc print data=Asia_only;
  run;
ods excel close;
```

Figure 9-14 – Hidden Columns Shown on the Output Excel Spreadsheet

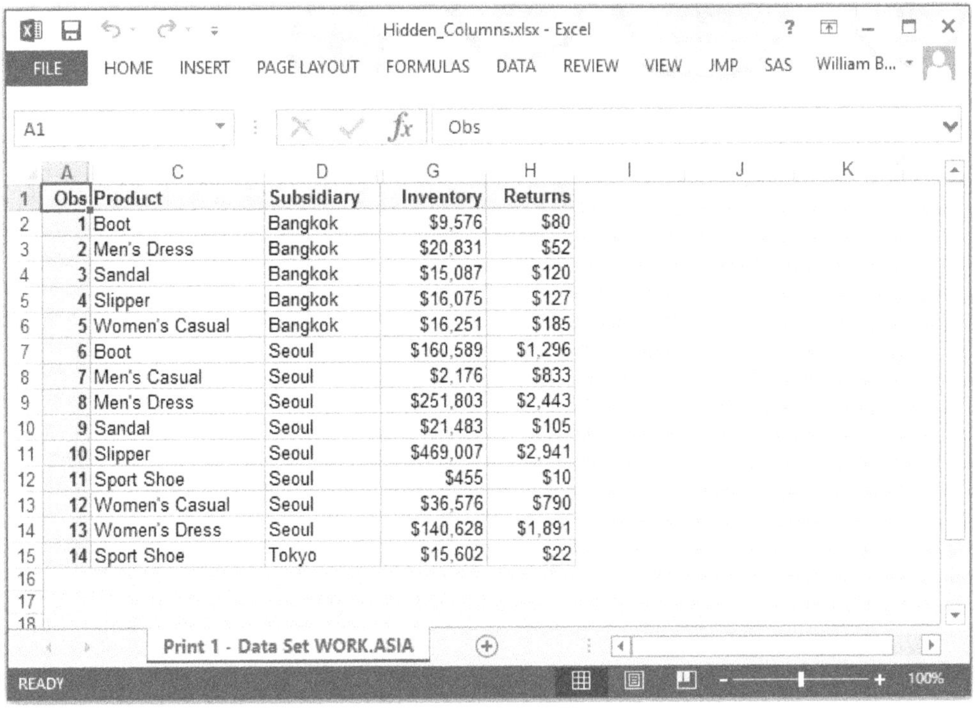

Obs	Product	Subsidiary	Inventory	Returns
1	Boot	Bangkok	$9,576	$80
2	Men's Dress	Bangkok	$20,831	$52
3	Sandal	Bangkok	$15,087	$120
4	Slipper	Bangkok	$16,075	$127
5	Women's Casual	Bangkok	$16,251	$185
6	Boot	Seoul	$160,589	$1,296
7	Men's Casual	Seoul	$2,176	$833
8	Men's Dress	Seoul	$251,803	$2,443
9	Sandal	Seoul	$21,483	$105
10	Slipper	Seoul	$469,007	$2,941
11	Sport Shoe	Seoul	$455	$10
12	Women's Casual	Seoul	$36,576	$790
13	Women's Dress	Seoul	$140,628	$1,891
14	Sport Shoe	Tokyo	$15,602	$22

Here columns 'B', 'E', and 'F' are hidden. Take note of the two vertical lines in the Excel header row between columns 'A' and 'C', and 'D' and 'G' the two vertical lines are close together and hard to see.

The ABSOLUTE_ROW_HEIGHT= Suboption

I always disliked opening an Excel workbook to adjust the heights of rows within the individual sheets. Now I use the ABSOLUTE_ROW_HEIGHT= option to perform that task for me when the sheets are created. The option string for this suboption uses the row heights list (separated by commas and measured in points) to adjust every row of the spreadsheet, including the first row of the spreadsheet. If only one row value exists in the string, then all rows are adjusted to that height. When more than one number is supplied, then the rows are given the individual heights in order of each number supplied. If more lines of data exist than numbers provided, then the pattern repeats itself until the data is exhausted. The default is that no special adjustments are made and all rows are the same height.

SAS Code 9-11 – SAS Code to Adjust Row Heights in a Repeating Pattern

```
ods excel file = "&path\Absolute_row_height.xlsx"
          options(ABSOLUTE_ROW_HEIGHT='15,30');
   proc print data=Asia_only;
  run;
ods excel close;
```

Figure 9-15 –ABSOLUTE_ROW_HEIGHT= Suboption Output for a Repeating Pattern

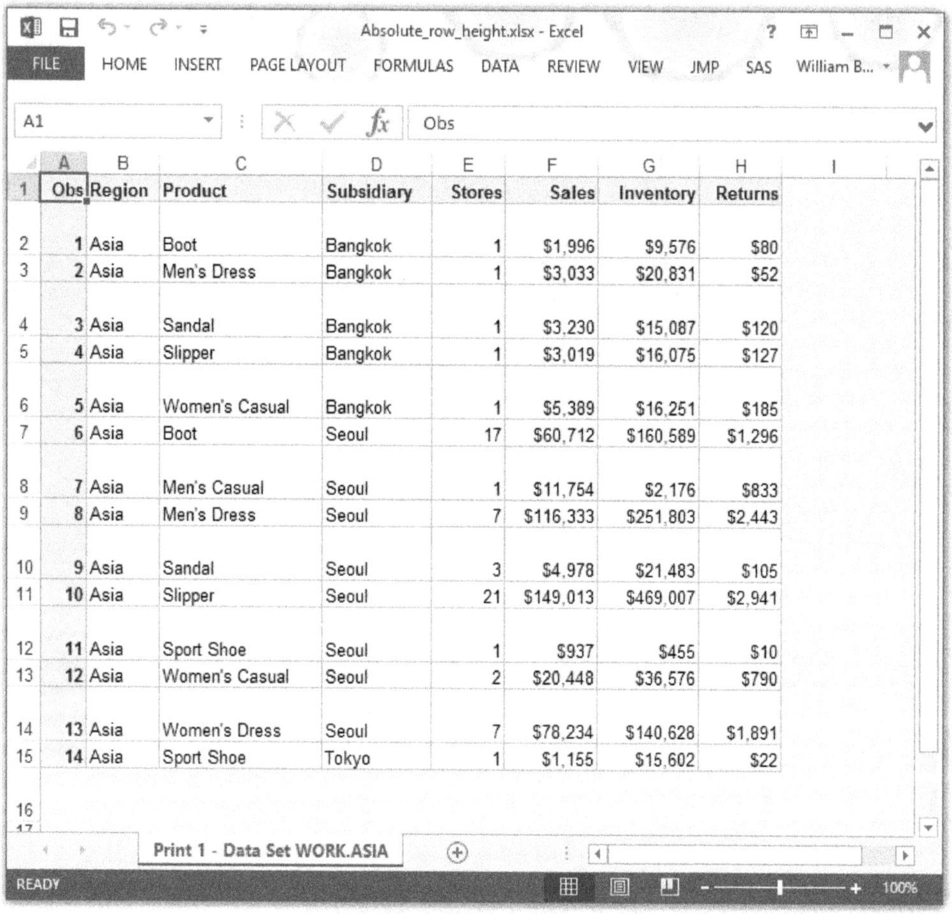

Obs	Region	Product	Subsidiary	Stores	Sales	Inventory	Returns
1	Asia	Boot	Bangkok	1	$1,996	$9,576	$80
2	Asia	Men's Dress	Bangkok	1	$3,033	$20,831	$52
3	Asia	Sandal	Bangkok	1	$3,230	$15,087	$120
4	Asia	Slipper	Bangkok	1	$3,019	$16,075	$127
5	Asia	Women's Casual	Bangkok	1	$5,389	$16,251	$185
6	Asia	Boot	Seoul	17	$60,712	$160,589	$1,296
7	Asia	Men's Casual	Seoul	1	$11,754	$2,176	$833
8	Asia	Men's Dress	Seoul	7	$116,333	$251,803	$2,443
9	Asia	Sandal	Seoul	3	$4,978	$21,483	$105
10	Asia	Slipper	Seoul	21	$149,013	$469,007	$2,941
11	Asia	Sport Shoe	Seoul	1	$937	$455	$10
12	Asia	Women's Casual	Seoul	2	$20,448	$36,576	$790
13	Asia	Women's Dress	Seoul	7	$78,234	$140,628	$1,891
14	Asia	Sport Shoe	Tokyo	1	$1,155	$15,602	$22

Here every other row is a different size. This repeating pattern continues until there is no more data to output.

The FROZEN_ROWHEADERS= Suboption

The ODS Excel FROZEN_ROWHEADERS= suboption is similar to the FROZEN_HEADERS= suboption, but instead does its work on the data in the rows. It sets up the left-most columns of the Excel worksheet so that they do not scroll, which is helpful for wide tables. The choices are either all columns scrolling (the default) or one or more columns frozen at the left of the worksheet. The values "ON" or the number "1" freeze only the left-most column, any other number freezes by that many columns. Note that other SAS procedures might produce output differently than the PRINT procedure.

SAS Code 9-12 – The FROZEN_ROWHEADERS= Default Option Value

```
ods excel file = "&path\Frozen_rowHeaders_on.xlsx"
         options(Frozen_rowHeaders='on');
  proc print data=Asia_only;
  run;
ods excel close;
```

Figure 9-16 – FROZEN_ROWHEADERS= 'ON' Showing Column 'B' Missing

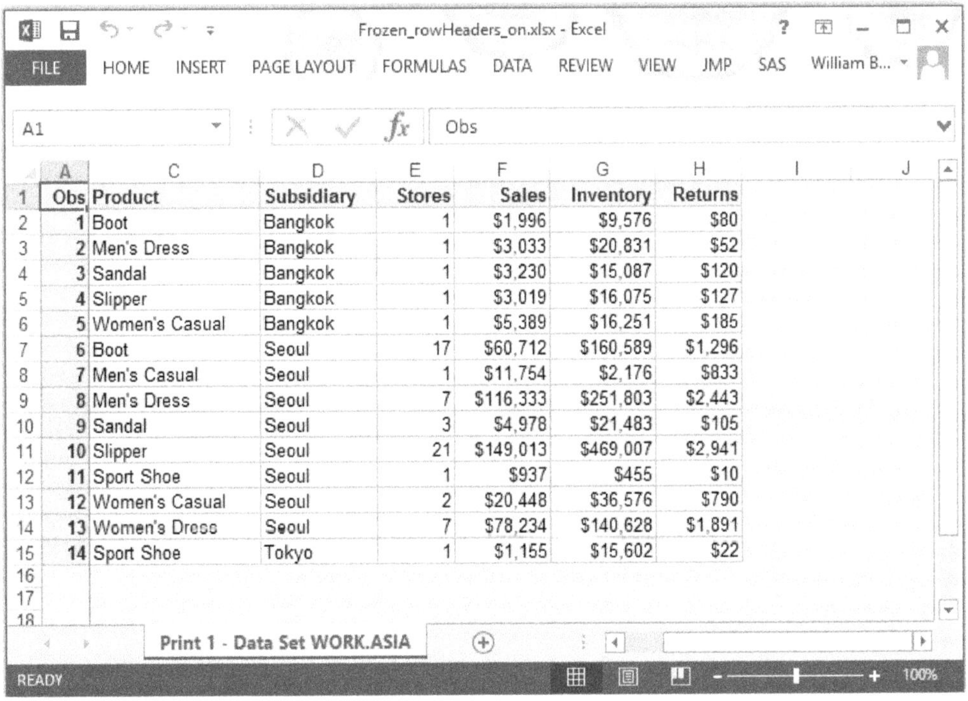

Obs	Product	Subsidiary	Stores	Sales	Inventory	Returns
1	Boot	Bangkok	1	$1,996	$9,576	$80
2	Men's Dress	Bangkok	1	$3,033	$20,831	$52
3	Sandal	Bangkok	1	$3,230	$15,087	$120
4	Slipper	Bangkok	1	$3,019	$16,075	$127
5	Women's Casual	Bangkok	1	$5,389	$16,251	$185
6	Boot	Seoul	17	$60,712	$160,589	$1,296
7	Men's Casual	Seoul	1	$11,754	$2,176	$833
8	Men's Dress	Seoul	7	$116,333	$251,803	$2,443
9	Sandal	Seoul	3	$4,978	$21,483	$105
10	Slipper	Seoul	21	$149,013	$469,007	$2,941
11	Sport Shoe	Seoul	1	$937	$455	$10
12	Women's Casual	Seoul	2	$20,448	$36,576	$790
13	Women's Dress	Seoul	7	$78,234	$140,628	$1,891
14	Sport Shoe	Tokyo	1	$1,155	$15,602	$22

Here I have scrolled the Excel worksheet data to the left (hiding column "B") by one column. This is shown by the fact that column 'B' is not visible. The other way to determine that the row header is frozen is the vertical line between columns 'A' and 'C'. The typical output from SAS does not show grid lines outside of the data output by SAS. The line is vertical between all rows but it is more visible below Excel rows 16 to 18 because the rest of the grid is missing.

SAS Code 9-13 – FROZEN_ROWHEADERS= Suboption Freezing Four Columns

```
ods excel file = "&path\Frozen_rowHeaders_number.xlsx"
         options(Frozen_rowHeaders='4');
  proc print data=Asia_only;
  run;
ods excel close;
```

Figure 9-17 - FROZEN_ROWHEADERS= with Four Columns Frozen

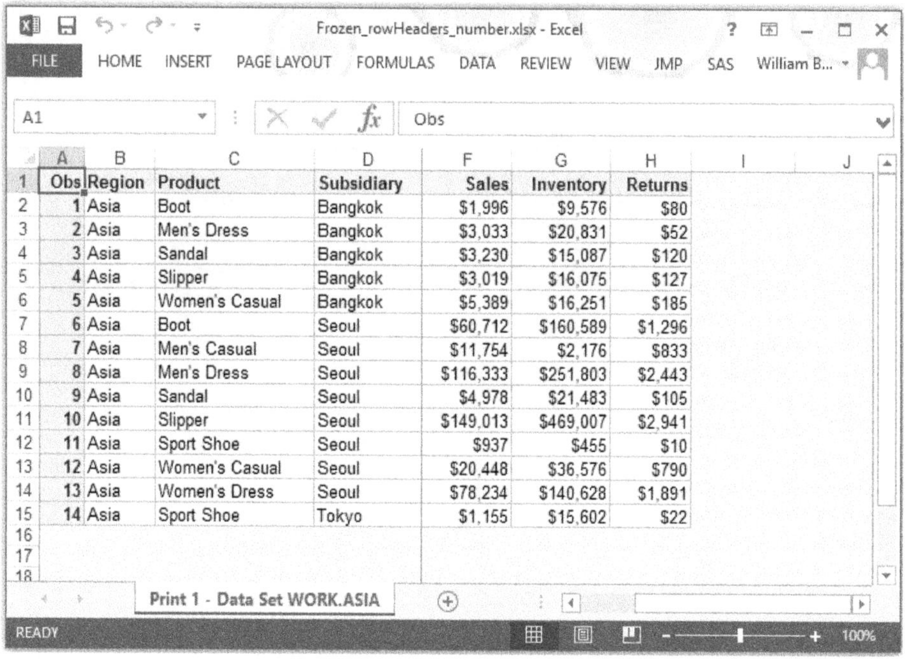

Obs	Region	Product	Subsidiary	Sales	Inventory	Returns
1	Asia	Boot	Bangkok	$1,996	$9,576	$80
2	Asia	Men's Dress	Bangkok	$3,033	$20,831	$52
3	Asia	Sandal	Bangkok	$3,230	$15,087	$120
4	Asia	Slipper	Bangkok	$3,019	$16,075	$127
5	Asia	Women's Casual	Bangkok	$5,389	$16,251	$185
6	Asia	Boot	Seoul	$60,712	$160,589	$1,296
7	Asia	Men's Casual	Seoul	$11,754	$2,176	$833
8	Asia	Men's Dress	Seoul	$116,333	$251,803	$2,443
9	Asia	Sandal	Seoul	$4,978	$21,483	$105
10	Asia	Slipper	Seoul	$149,013	$469,007	$2,941
11	Asia	Sport Shoe	Seoul	$937	$455	$10
12	Asia	Women's Casual	Seoul	$20,448	$36,576	$790
13	Asia	Women's Dress	Seoul	$78,234	$140,628	$1,891
14	Asia	Sport Shoe	Tokyo	$1,155	$15,602	$22

In Figure 9-17 I have scrolled the Excel worksheet data to the left by one column, hiding column "E". The other way to determine that the row header is frozen is the vertical line between columns 'D' and 'F'. This output from SAS does not show grid lines outside of the cells A1 to H15. The vertical line indicating the frozen headers is vertical between all rows but it is more visible in Excel rows 16 to 18 because the rest of the grid is missing

The HIDDEN_ROWS= Suboption

Hiding rows in an Excel worksheet can now be done when SAS is executing instead of doing it manually after the workbooks are created. The HIDDEN_ROWS= suboption is flexible; you can combine individual rows and ranges in the same 'number_list_range' value string. The default option is to not have any rows hidden.

SAS Code 9-14 – The HIDDEN_ROWS= Suboption with a Row and a Range of Rows Hidden

```
ods excel file = "&path\Hidden_Rows.xlsx"
          options(Hidden_Rows='4,8-10');
   proc print data=Asia_only noobs;
  run;
ods excel close;
```

Figure 9-18 – Excel Output with Hidden Rows

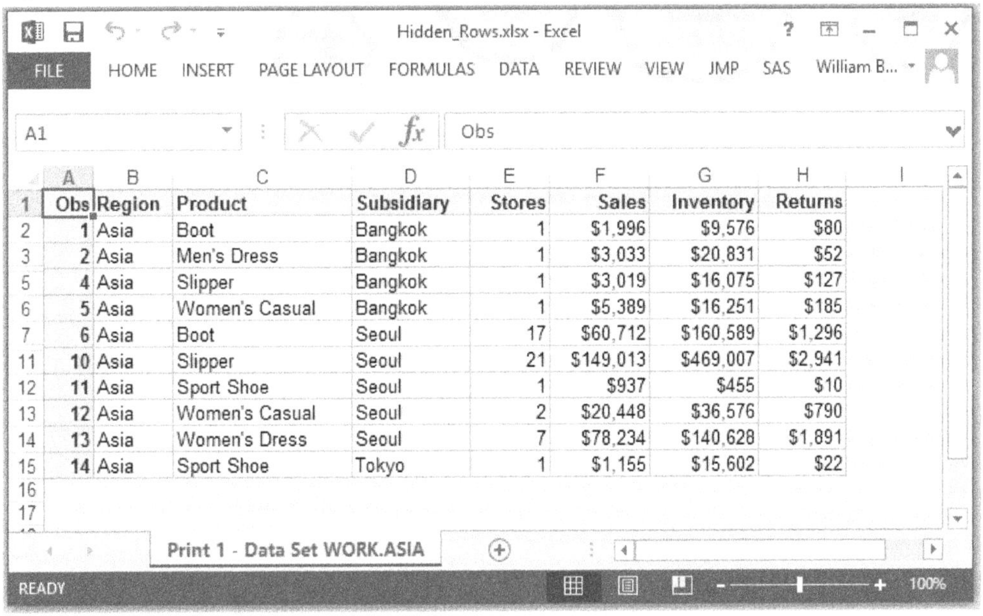

	A	B	C	D	E	F	G	H
1	Obs	Region	Product	Subsidiary	Stores	Sales	Inventory	Returns
2	1	Asia	Boot	Bangkok	1	$1,996	$9,576	$80
3	2	Asia	Men's Dress	Bangkok	1	$3,033	$20,831	$52
5	4	Asia	Slipper	Bangkok	1	$3,019	$16,075	$127
6	5	Asia	Women's Casual	Bangkok	1	$5,389	$16,251	$185
7	6	Asia	Boot	Seoul	17	$60,712	$160,589	$1,296
11	10	Asia	Slipper	Seoul	21	$149,013	$469,007	$2,941
12	11	Asia	Sport Shoe	Seoul	1	$937	$455	$10
13	12	Asia	Women's Casual	Seoul	2	$20,448	$36,576	$790
14	13	Asia	Women's Dress	Seoul	7	$78,234	$140,628	$1,891
15	14	Asia	Sport Shoe	Tokyo	1	$1,155	$15,602	$22
16								
17								

Figure 9-18 shows that row 4 and rows 8, 9, and 10 are hidden. Notice the double line in the Excel header column between rows 3 and 5 and also between rows 7 and 11. Make sure that you do not confuse the Excel row header column with the SAS OBS number, which also skips.

The ROW_HEIGHTS= Suboption

The 'HEIGHTS' part of the suboption name ROW_HEIGHTS is plural, and as the plural name implies, the suboption has an affect on several row heights. A single value of zero indicates that the font size should be used as the height of all of the row types described in the list below. For this option you need to provide a quoted string of positional parameters to identify the seven row heights listed below. Values can be skipped but skipped parameters require a comma to show that the parameter is missing, except the last used value. As mentioned above, row heights are measured in points.

- Table header rows
- Table body rows
- BY value lines
- Titles
- Footers
- Page break height
- Paragraph skip height

SAS Code 9-15 – Code Options to Display Row Heights Associated with the ROW_HEIGHTS= Option

```
ods excel file = "&path\Row_Heights.xlsx"
          options(ROW_HEIGHTS='20,15,15,35,55,25,5'
                   embedded_titles='on'
                   embedded_footnotes='on'
                   sheet_interval='none');
Title    'my title';
footnote 'my footnote';
  proc print data=asia_only;
  by Subsidiary;
  run;
ods excel close;
```

Figure 9-19 – Excel Output Showing the Different Row Heights Produced by SAS Code 9-15

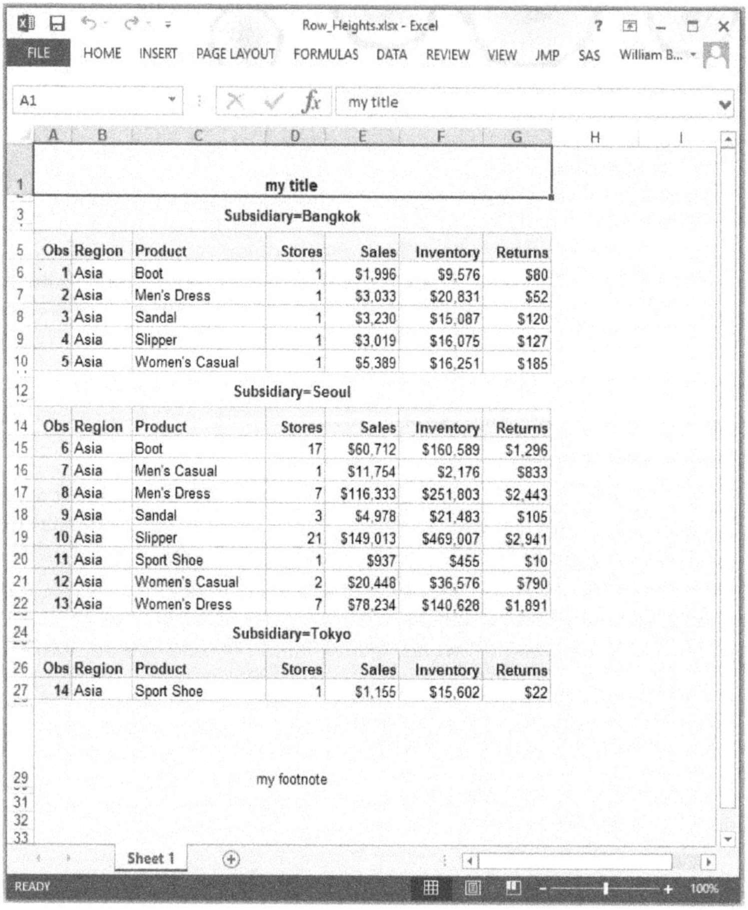

my title

Subsidiary=Bangkok

Obs	Region	Product	Stores	Sales	Inventory	Returns
1	Asia	Boot	1	$1,996	$9,576	$80
2	Asia	Men's Dress	1	$3,033	$20,831	$52
3	Asia	Sandal	1	$3,230	$15,087	$120
4	Asia	Slipper	1	$3,019	$16,075	$127
5	Asia	Women's Casual	1	$5,389	$16,251	$185

Subsidiary=Seoul

Obs	Region	Product	Stores	Sales	Inventory	Returns
6	Asia	Boot	17	$60,712	$160,589	$1,296
7	Asia	Men's Casual	1	$11,754	$2,176	$833
8	Asia	Men's Dress	7	$116,333	$251,803	$2,443
9	Asia	Sandal	3	$4,978	$21,483	$105
10	Asia	Slipper	21	$149,013	$469,007	$2,941
11	Asia	Sport Shoe	1	$937	$455	$10
12	Asia	Women's Casual	2	$20,448	$36,576	$790
13	Asia	Women's Dress	7	$78,234	$140,628	$1,891

Subsidiary=Tokyo

Obs	Region	Product	Stores	Sales	Inventory	Returns
14	Asia	Sport Shoe	1	$1,155	$15,602	$22

my footnote

In this example, I used the BYGROUP (Subsidiary) to split the data into segments that would show different types of spacing used in the Excel output. Titles and footnotes are larger than the other lines. Titles is highlighted in green because it includes cell "A1", which is the default curser position when Excel workbooks are first opened. Rows '2', '4', '11', '13', '23', '25', '28', and '30' are so small that they are nearly hidden. The subsidiary header lines are slightly larger than the data lines and the BY lines are surrounded by very small lines.

The ROW_REPEAT= Suboption

The results of ROW_REPEAT= suboption are harder to identify than the COLUMN_REPEAT= output. If your data does not extend to a second page, then the suboption is useless. This suboption allows copying information from the middle of page one or a header onto subsequent sheets of a long printout. I am using the SASHELP.ORSALES SAS data set rows 1 to 63. The first and second "Print Preview" page output is displayed below in Figure 9-21 and Figure 9-22. Valid options include 'NONE', a 'number', a 'number-range', or 'HEADER' to describe what you want to repeat. Also, using the number '1' is the same as using 'HEADER', Neither FROZEN_HEADERS= nor FROZEN_ROWHEADERS= affect the number of header columns.

SAS Code 9-16 – Code to Output Repeated Row Values One Row

```
ods excel file = "&path\ROW_REPEAT_number.xlsx"
          options(SHEET_INTERVAL='none'
                  ROW_REPEAT='6');
  proc print data=SASHELP.ORSALES(obs=63);
  run;
ods excel close;
```

Figure 9-20 – Page Setup Sheet Showing Row 6 Repeated on Top of Page 2

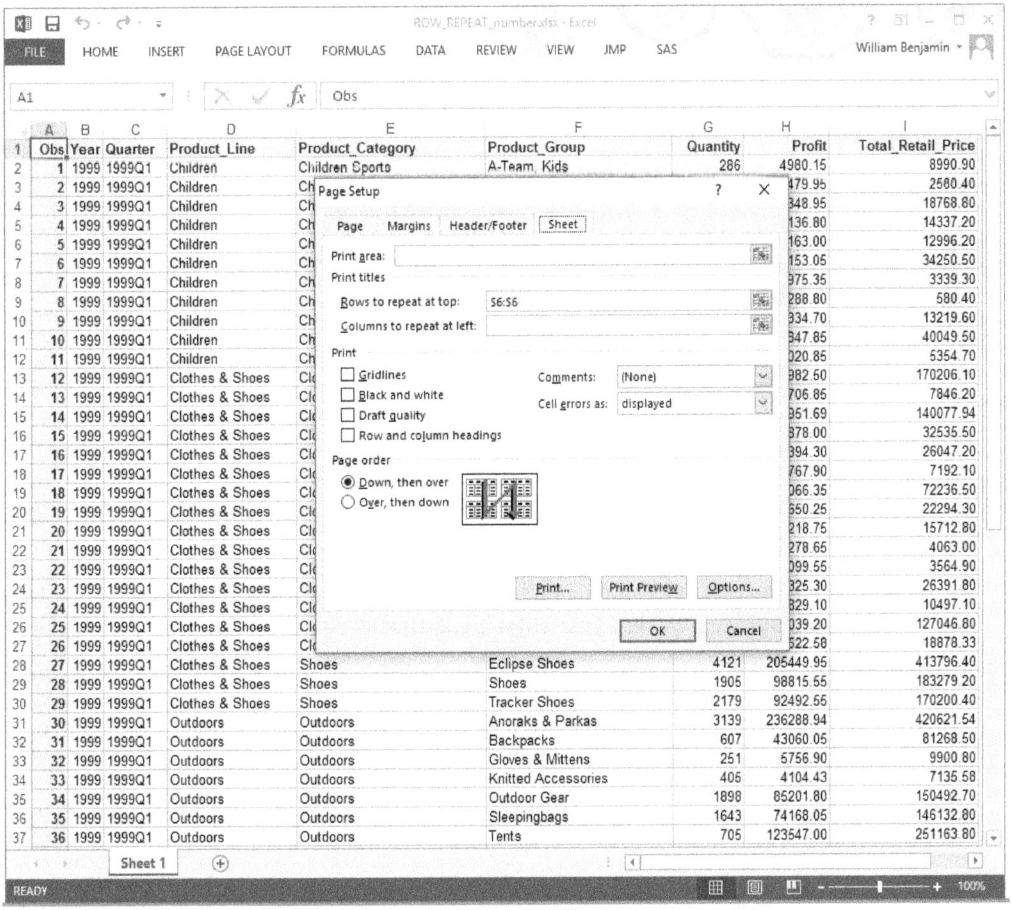

Again the '$' is Excel notation for the absolute row number '6'.

Figure 9-21 – ROW_REPEAT= Output Page 1

Figure 9-21 shows the first page of the Print Preview sheet. Because the width of the output exceeds the width of a portrait formatted page, four sheets are output in "Down Then Over" page order. I am only showing the first two pages. Also, because the 'Page Setup' sheet obscures some of the data, it is difficult to confirm exactly which row of data is duplicated. You need to be careful because the row that is duplicated is the 6th row of the Excel worksheet, not the 6th SAS observation.

In Figure 9-22 the top row of the output on the Print preview sheet is Excel row '6' from the first page of the output in Figure 9-21. Admittedly, it is difficult to see on either a printed page or mobile device.

Figure 9-22 – ROW_REPEAT= Output Second Half

Be careful, because the row that is duplicated is the 6th row of the Excel worksheet, not the 6th SAS observation.

The code in SAS Code 9-17 duplicates a range of rows. The value '3-5' duplicates the third, fourth, and fifth row of the Excel output worksheet onto the second and subsequent output pages, not SAS observations.

SAS Code 9-17 – Code to Output Repeated Range of Row Values

```
ods excel file = "&path\ROW_REPEAT_range.xlsx"
         options(SHEET_INTERVAL='none'
               ROW_REPEAT='3-5');
  proc print data=SASHELP.ORSALES(obs=63);
  run;
ods excel close;
```

Figure 9-23 – The ROW_REPEAT= Suboption Using a Range

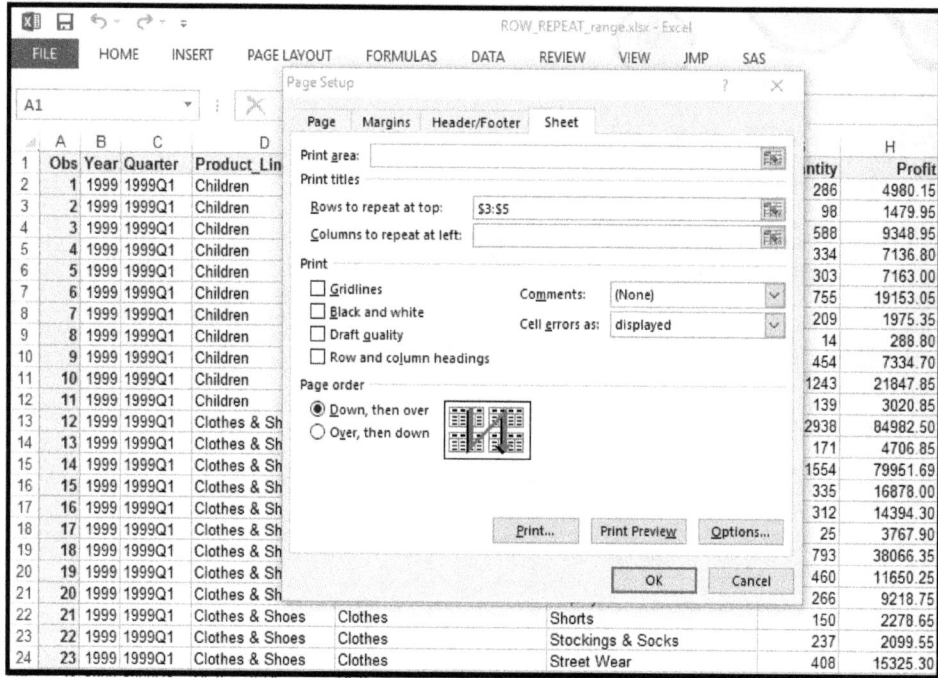

Rows 3, 4, and 5 are repeated at the top of the page after page 1.

Figure 9-24 – ROW_REPEAT= Output Using a Range

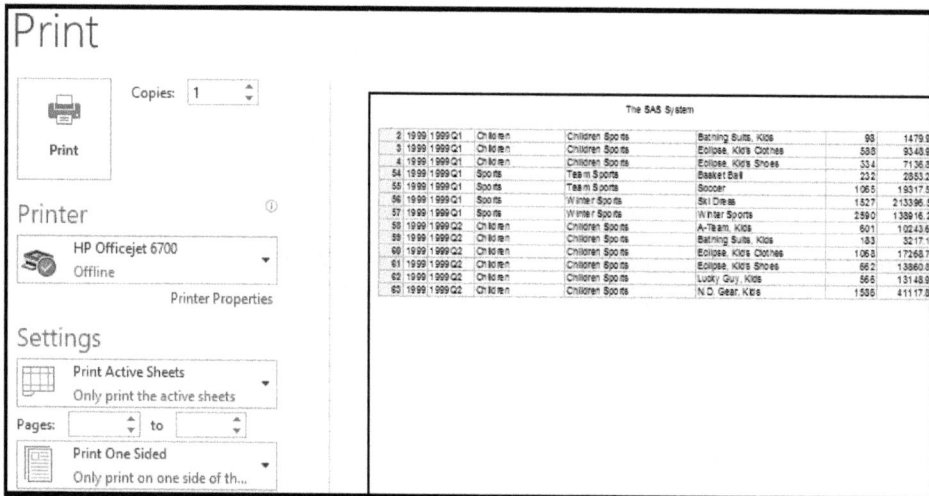

Excel observations 2, 3, and 4 appear at the top of the printout. This data is from the SAS data set rows 3, 4, and 5, because you include the headers in your count. Only part of the page is reproduced in Figure 9-24.

SAS Code 9-18 – Code to Output Repeated Header Row Values

```
ods excel file = "&path\ROW_REPEAT_header.xlsx"
          options(SHEET_INTERVAL='none'
                  ROW_REPEAT='header');
  proc print data=SASHELP.ORSALES(obs=63);
  run;
ods excel close;
```

SAS Code 9-18 reproduces the header from the first page on all of the output pages.

Figure 9-25 – Showing the ROW_REPEAT='HEADER' Option Settings

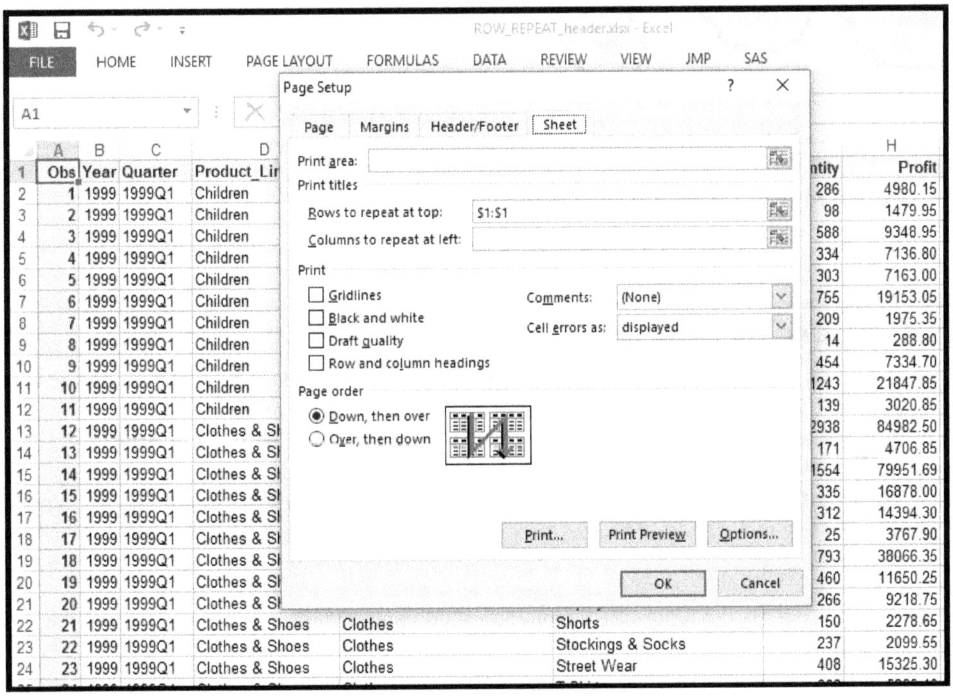

Only row one of the spreadsheet is repeated on the pages.

Figure 9-26 – Output from the ROW_REPEAT= 'HEADER' Setting

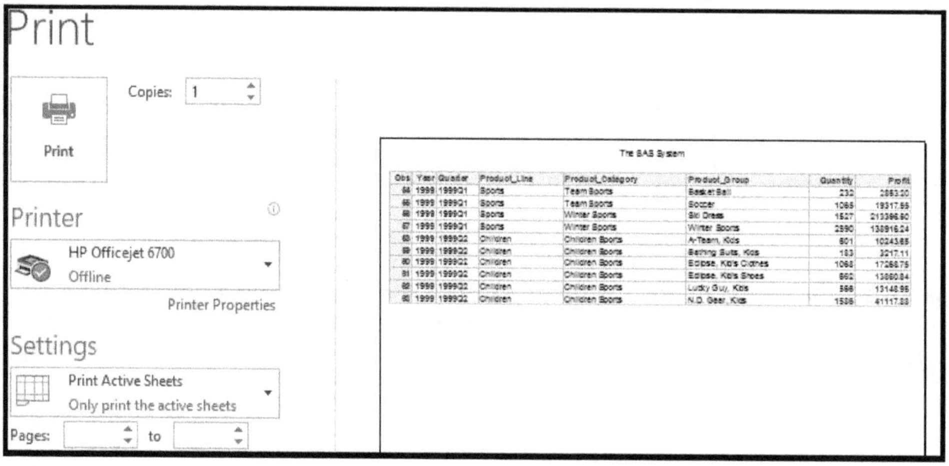

This is another way to get the headers to print on the output pages. Only part of the page is reproduced in Figure 9-26.

The FORMULAS= Suboption

Anyone who has used Excel for any length of time has probably become familiar with the fact that you can use Excel formulas to calculate any number of things. Well, this suboption enables you to use SAS to set up formulas as fields in the output data set to be implemented when Excel opens the data file. The default is that using the data as a formula is turned off. When you want to use the FORMULAS='ON' suboption, any data field that begins with an equal sign (=) activates the formula processing. The FORMULAS= suboption might not be the only way to add a formula to an Excel worksheet.

The DATA step here is creating a SAS character value that contains an equal sign, and the values of the fields Sales and Returns so that the net value of Sales minus Returns can be calculated. When formula processing is turned off, the character values are visible. When formulas are turned on, the calculation occurs and the value is displayed because of the prepended equal sign.

The code in SAS Code 9-19 is required for use with both SAS Code 9-20 and SAS Code 9-21. This code builds a SAS data set that has a new column called "NET_SALES". The formula subtracts returns from sales to calculate "NET_SALES". This calculation is not done for the header row by adding one to the value of the _n_ variable. Other methods of reading the ASIA_ONLY data set might not work the same as the method I choose here.

SAS Code 9-19 Example of how to setup a data field to activate Excel formula processing.

```
data ASIA_FORMULAS;
  set Asia_only;
  row = _n_ + 1;
  net_sales =   '='                            ||
                left(trim(put(sales,9.)))      ||
                '_'                            ||
                left(trim(put(returns,9.)))
                ;
run;
```

Execute the code in SAS Code 9-19 before using either SAS Code 9-20 or SAS Code 9-21. No output figure is shown for this code.

SAS Code 9-20 Print Example with the FORMULA sub-option set to 'OFF'.

```
ods excel file = "&path\Formulas_off.xlsx"
          options(formulas='off');
  proc print data=Asia_formulas;
  run;
ods excel close;
```

Figure 9-27 Net_Sales shown with formulas turned off.

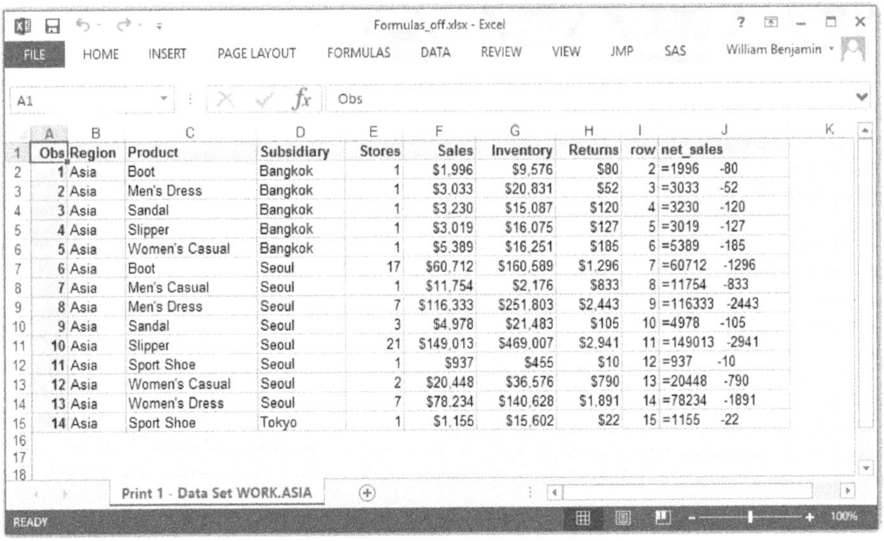

Without formulas turned on, the Net_Sales column does not make sense.

SAS Code 9-21 – Print Example with the FORMULA= Suboption Set to 'ON'

```
ods excel file = "&path\Formulas_on.xlsx"
          options(formulas='on');
  proc print data=Asia_formulas;
  run;
ods excel close;
```

Figure 9-28 – Net_Sales Shown with Formulas Turned On

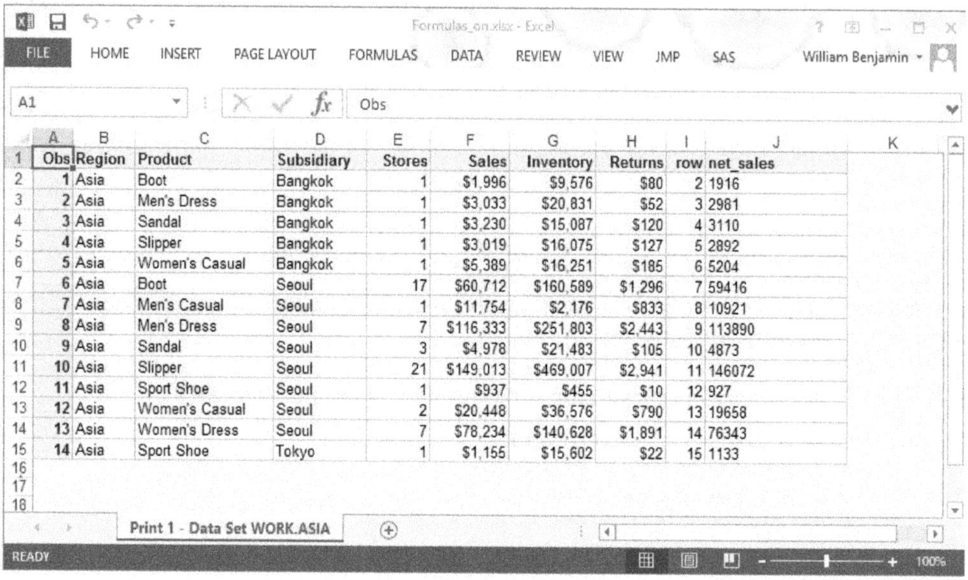

This image shows that the formula for Net_Sales was executed. But notice that the column is left justified, which indicates that the field contains a character value and, by default, is not formatted.

The START_AT= Suboption

The START_AT= suboption is a handy way to place your output into the Excel worksheet somewhere other than the default top-left corner of the worksheet (cell A1). This option gives you the freedom to place the output into any row and column of your output worksheet, allowing you to move your data under rows of blank cells. CSS style sheets described in Chapter 6 can be used to add images into the worksheet, and the START_AT= suboption can be used to position your data. SAS 9.4M4 also has added the ability to use alphabetic characters for the column identifiers as in cell "A1". The value '4,5' is 'Column,Row' format.

SAS Code 9-22 – Using the START_AT= Suboption to Offset the Data within the Excel Worksheet

```
ods excel file = "&path\Starting_at.xlsx"
        options(start_at='4,5');
  proc print data=Asia_only;
  run;
ods excel close;
```

Figure 9-29 – Showing the Output Data Offset to cell 'D5', Column 4, Row 5

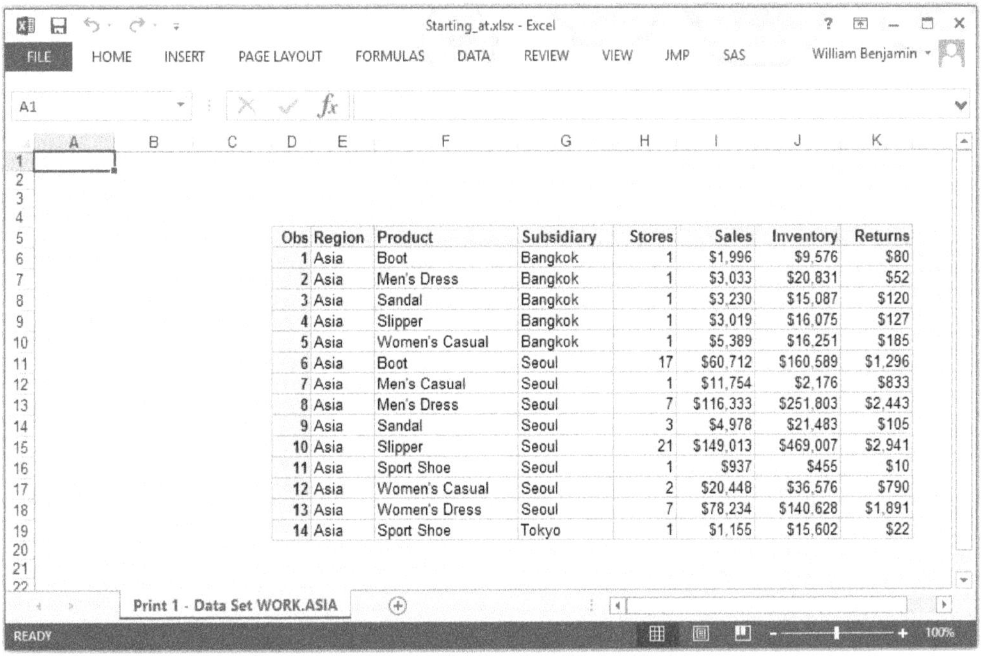

Obs	Region	Product	Subsidiary	Stores	Sales	Inventory	Returns
1	Asia	Boot	Bangkok	1	$1,996	$9,576	$80
2	Asia	Men's Dress	Bangkok	1	$3,033	$20,831	$52
3	Asia	Sandal	Bangkok	1	$3,230	$15,087	$120
4	Asia	Slipper	Bangkok	1	$3,019	$16,075	$127
5	Asia	Women's Casual	Bangkok	1	$5,389	$16,251	$185
6	Asia	Boot	Seoul	17	$60,712	$160,589	$1,296
7	Asia	Men's Casual	Seoul	1	$11,754	$2,176	$833
8	Asia	Men's Dress	Seoul	7	$116,333	$251,803	$2,443
9	Asia	Sandal	Seoul	3	$4,978	$21,483	$105
10	Asia	Slipper	Seoul	21	$149,013	$469,007	$2,941
11	Asia	Sport Shoe	Seoul	1	$937	$455	$10
12	Asia	Women's Casual	Seoul	2	$20,448	$36,576	$790
13	Asia	Women's Dress	Seoul	7	$78,234	$140,628	$1,891
14	Asia	Sport Shoe	Tokyo	1	$1,155	$15,602	$22

Print 1 - Data Set WORK.ASIA

Often a company will have a logo used specifically to brand corporate Excel spreadsheets. This option gives you a way to free up space for that logo (or anything else) without having to manually move the data. See Chapter 6 for using CSS style options to add images.

Conclusion

In this chapter, we discussed many ODS Excel destination suboptions that allowed us to manipulate features of the Excel worksheet rows, columns, and cells at the time that SAS is executing. With these options, you are able to set filters, hide columns or rows, freeze headers or row labels, adjust the height and width of the output rows, execute formulas, and even control the spacing of the whole report. These options are especially advantageous for anyone who does not want to open their Excel workbooks to modify the sheets.

Index

W

WC3/CSS3 (website) 53
WORK= option 8
workbooks, options affecting 31–44
"Working with the SAS® ODS EXCEL Destination to
          Send Graphs, and Use Cascading Style
          Sheets When Writing to EXCEL
          Workbooks" (Benjamin) 54
worksheets
    arguments affecting worksheet-level output
        features 46
    EMBED_TITLES_ONCE= option with multiple
        84–85
    EMBED_TITLES_ONCE= option with one 83–
        84
    options affecting features of 69–107

X

*.xls file structure 2
*.xlsx file structure 2
XML file 5

Z

ZIP file 5
ZOOM= option 9, 33, 43–44

Ready to take your SAS® and JMP® skills up a notch?

Be among the first to know about new books, special events, and exclusive discounts.
support.sas.com/newbooks

Share your expertise. Write a book with SAS.
support.sas.com/publish

sas.com/books
for additional books and resources.

§.sas.

THE POWER TO KNOW®

CPSIA information can be obtained
at www.ICGtesting.com
Printed in the USA
BVHW07s2203220518
517070BV00003B/82/P

9 781629 606095